WILDLIFE PARADISES

WILDLIFE PARADISES

A WORLDWIDE GUIDE **BY JOHN GOODERS**

with a foreword by Roger Tory Peterson

Praeger Publishers
New York

Published in the United States of America in 1975
by Praeger Publishers, Inc., 111 Fourth Avenue, New York, N.Y. 10003

© London Editions Ltd. 1975

Library of Congress Cataloging in Publication Data
Gooders, John.
  Wildlife paradises.
  1. Wildlife refuges.   2. Zoology.   I. Title.
  QH75.G64 1975   599   74-5581
ISBN 0-275-22140-7

D.L.: M. 22.005-1975.   Printed in Spain

(*Previous page*) Scarlet ibises, the most
startlingly plumaged birds of the
South American swamps, are only easily
to be seen today in Trinidad

Location of wildlife paradises

Figures refer to the numbers to
the left of the Contents (opposite)

Contents

Foreword by Roger Tory Peterson

Thumbing through the pages of this exciting book, I discover that of the fifty-eight wildlife paradises listed by John Gooders as the greatest in the World, I have visited thirty-five — three-fifths of the lot. I am determined to eventually see them all.

Some years ago I made the pronouncement that the flamingoes of Lake Nakuru presented the greatest bird spectacle on Earth. I still hold to that opinion — but since I made that statement I have seen the Chinchas of Peru with their millions of guanay cormorants, piqueros and pelicans; I have also wandered amongst the vast colonies of Adélie penguins and emperors at Cape Crozier in the Antarctic. In their busy, noisy multitudes, they are every bit as impressive, lacking only the rosy pink of a million flamingoes.

Nakuru, the Chinchas and Crozier are all refuges of a sort and wildlife would seem to be secure for the foreseeable future. Safe, too, are many of the other paradises described in this book, but not all. It will be increasingly difficult to maintain the integrity of some of these places, if we should create or perpetuate the social, economic, political, chemical or moral climate that proves inimical to their survival.

And what of the mammals? They are even more vulnerable than the birds. Will our grandchildren still be able to see a half million wildebeest and zebra in migration on the Serengeti as my wife and I observed them one recent April? I hope that I shall never see the herds of the Serengeti reduced to the last few survivors as is the predicament of the Walia ibex in the Simien Mountains of Ethiopia, another paradise listed in this book. Paradise it is, with canyons to match the Grand Canyon, but a sort of Paradise Lost. I did manage to see several ibex, but almost at the limit of my vision, and on slopes so precipitous that the local tribesmen could not negotiate them to burn the vegetation and plant their barley. But in a land that is losing out to starvation, how can such an area be saved — even though it be named a park or reserve? This is a dilemma that may become acute in many other places in the decades ahead.

But looking at the positive side: there has been a great surge of interest in wildlife, a breakthrough in environmental awareness on the part of the public, that we could not have anticipated a few years ago. The pressures on wildlife have generated their counterpressures, and with luck the whooping cranes will always fly between Wood Buffalo Park in Canada and Aransas National Wildlife Refuge in Texas; Barro Colorado in the Canal Zone will maintain its tropical ecosystem intact; Carlsbad Caverns will still be the roosting place of millions of bats; Caroni Swamp in Trinidad still delight the tourist with the evening flight of thousands of scarlet ibis; and the Galapagos will remain an unparalleled theater of evolution.

What worries me, however, is that thousands of lesser paradises may go by the board because they are not publicized enough. It is easier to save a place when it is number one – the best place for flamingoes – or antelope – or whatever.

I am very much impressed with the selection that John Gooders has made of 'the greatest'. He has not missed one that I would have included, except the Bear River Refuge in Utah which in its way is the equal of the Camargue in France, the Coto Doñana in Spain or the Aransas in Texas. On the other hand he has included several unique places of which I have had no knowledge. These I must visit.

My dictionary defines *Eden* as 'The garden that was the first home of Adam and Eve: often called *Paradise*; hence any delightful region or abode.' My own personal definition of Eden is a bit different: a place that remains forever wild, unspoiled by man or not too much affected by his presence.

My personal Edens fall into two basic categories: benign Edens (as was the Garden of Eden before it was violated by Adam and Eve) and harsh Edens, that remain much as they are because they are unsuited for human occupancy. The Hawaiian Islands were certainly a benign Eden when the first Polynesian settlers arrived, so benign that they built up a population of hundreds of thousands. The water-deficient Galapagos are a prime example of a harsh Eden. They harbored no endemic race of men when the first Europeans stepped ashore and to this day there are only three or four small communities in the sprawling archipelago.

Some of the paradises described in this book are true Edens – pristine, unchanged for centuries except perhaps for a certain amount of ecological succession and a negligible amount of human disturbance. The Antarctic, Barro Colorado in Panama, Caroni Swamp in Trinidad, some of the Galapagos Islands (except for the depletion of tortoises), Lake Nakuru in Africa, and Wood Buffalo Park in Alberta are representative. Other paradises have been managed by biologists to create a maximum carrying capacity for wildlife. The Aransas in Texas, Klamath in California, The Chinchas in Peru, are good examples. Still others are subject to heavy human visitation, yet manage to maintain their wildlife populations intact. Ngorongoro Crater and Nairobi National Park in East Africa, Kruger Park in South Africa and Everglades National Park in Florida fall into this category.

In fact, there are any number of Federal and Provincial refuges in North America and elsewhere that might qualify as paradises, many of which are deliberately manipulated, by means of water impoundments and planting, to maintain maximum wildlife populations. In addition, every continent has a growing network of sanctuaries and natural parks. Some are little known to the general public, but it would take a shelf of books to detail all of them. In this single volume, John Gooders informs us selectively about 'the greatest'.

Introduction by John Gooders

Every so often wildlife enthusiasts come across a really outstanding area and are tempted to refer to it as a 'paradise'. With a little education and more experience they are inclined to revalue their opinion. The area might be 'great', 'first-class', 'A1' or even just 'OK', but the term 'paradise' has definite implications that it is very difficult to live up to. Almost by definition there cannot be very many paradises around. It is the exceptional sites that maintain their status while other places, equally exciting at the time, fade from the memory. To choose a paradise is a process of distillation—only the most outstanding areas in the world of wildlife can maintain their places. Some such places are presented in this book.

In the following pages outstanding wildlife areas are described and detailed as are their most important or interesting inhabitants. It is often difficult to decide exactly which are the most important animals in a reserve or refuge—it is a matter of judgement and I shall not be surprised if some people disagree with my choice. Some areas are paradises because of the extreme attraction or rarity of one or two of their inhabitants, others because they offer such a wealth of different species; and others still because a single species creates a pure wildlife spectacle. Lake Nakuru in Kenya is a place for flamingoes and waterbirds—but it is surely a paradise. Chitawan in Nepal has only two hundred rhinos—but the great Indian one-horned rhinoceros is a wildlife spectacle on its own!

At a time when we are becoming increasingly conscious of what man is doing to this planet, when lobbyists talk and argue incessantly about pollution and ecology, it is not surprising that more attention should be given to the voices that have been crying in the wilderness for so long, crying for help for our endangered wildlife. Most of us now believe that wildlife should be saved—we believe, though we have not yet persuaded our governments to allocate the resources necessary to do very much about it. We believe that the tiger should be saved but we are not prepared to dip into our pockets to do so. We must face the fact that the tiger, and all other wildlife, needs living space; that it's the land rather than the individual animal that is crucial to the endeavour. Habitat, and particularly paradise habitat, must be saved and maintained and protected. This is a newer more positive and constructive approach. Perhaps it will come too late for the tiger for it is difficult to grow a jungle to order;

but we have learned to create marshes for birds. Perhaps we can learn to create other habitats for other animals.

The paradises in this book are among the most important wildlife areas on earth. The celebrated Red Data Book of the Survival Service Commission of the International Union for the Conservation of Nature (IUCN) lists animals in danger. The World's Wildlife Paradises lists some of the habitats in danger. Save these areas and wildlife may survive, but even as this guide is written, habitats described are being destroyed. There may be a hope—but it is a thin one.

John Gooders, London 1974

Antarctica

The Antarctic is one huge great refuge, a
wildlife paradise on the grand scale.
And yet most of this vast continent is
covered with barren ice. Though Captain
Scott spotted a southern skua within
180 miles of the South Pole, wildlife is
confined to the perimeter of the continent
where the land meets the sea that is the
foundation of all Antarctic life. Even so the
length of coastline of Antarctica is
enormous and impossible for any
individual to explore on his own.
Fortunately wildlife travel agents are
becoming progressively more ambitious;
indeed there is hardly a place that some
agent has not thought of exploring, and
the Antarctic is no exception.

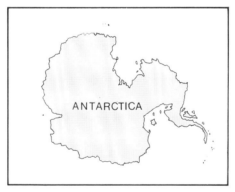

ANTARCTICA

   Tours usually concentrate on the
Antarctic Peninsular that extends northwards towards the southern tip of South
America. Here lie the South Orkneys and
the South Shetlands, King George Island,
Paradise Bay and, perhaps the most
famous of them all, Deception Island.
For anyone who enjoys the history of
nineteenth-century exploration the cruise
(for that's about the only way that anyone
will get here) is sheer romance. The
wildlife is fantastic too.
   From the time that the Falkland Islands
are left behind and the ship enters the
Drake Strait there is a feeling of intense
anticipation. Soon there are wandering
albatrosses following in the wake, the
largest birds in the world with a wing
span of eleven feet that carries them

The relationship between Antarctic animals is seldom a peaceful one, but Weddell seals (left) and emperor penguins (below) seldom come into violent conflict. The penguins breed on the ice in the depths of the Antarctic winter and are thus the only birds that never set foot on land. Young seals must also endure some of the worst conditions on earth, but are insulated from the cold by a thick layer of blubber.

round and round the globe with the prevailing winds. Then there are whales, including the great blue whale (the largest creature that has ever existed), sperm whales and sometimes killer whales, though these are more common further south, off the pack-ice itself.

On King George Island at Admiralty Bay is a mixed penguin rookery containing gentoo, chinstrap and Adelie penguins. There are the huge elephant seals as well as southern skuas, kelp gulls, giant petrels and the all-white sheathbills.

Torgeson Island boasts huge rookeries of Adelie penguins, and on Adelaide Island there is a rookery of emperor penguins, one of the strangest birds in the world. It is the only bird that *never* comes to land, breeding in the middle of the Antarctic winter on the ice itself and resting its single egg on its flippers as insulation against the cold.

Here and there throughout the islands are rookeries of crabeater, Weddell and elephant seals and of the notorious leopard seal. This voracious feeder lives on a diet of penguins and fish and has only the killer whale to fear in these waters. The smaller birds include snow petrels, and the graceful Wilson's petrels that patter with their feet over the surface of the sea as they bend to pick food from the water. There are also sooty and black-browed albatrosses to be seen.

Visiting: It is perhaps fortunate that Antarctica is so inaccessible, for its wildlife can thus be preserved. Lars-Eric Lindblad runs the MS *Lindblad Explorer* to Antarctica every year.

Species of particular interest

Blue Whale
  Balaenoptera musculus
Sperm Whale
  Physeter catodon
Killer Whale
  Orcinus orca
Crabeater Seal
  Lobodon carcinophagus
Weddell Seal
  Leptonychotes weddelli
Elephant Seal
  Mirounga angustirostris
Leopard Seal
  Hydrurga leptonyx
Wandering Albatross
  Diomedea exulans
Black-browed Albatross
  Diomedea melanophris
Emperor Penguin
  Aptenodytes fosteri
Adelie Penguin
  Pygoscelis adeliae
Chinstrap Penguin
  Pygoscelis antarctica
Gentoo Penguin
  Pygoscelis papua
Southern Skua
  Catharacta skua
Wilson's Petrel
  Oceanites oceanicus
Sheathbill
  Chionis alba

Among the numerous Antarctic birds,
Adélie penguins (below) gather at huge
rookeries where their elaborate greeting
ceremonies and aggressiveness have an
instant appeal to the visitor. King penguins
(left), though only slightly shorter than the
emperor penguins, weigh half as much.
Like their larger relatives they have no nest
territories to defend and are therefore
less quarrelsome than the Adélies.

Aransas National Wildlife Refuge, Texas, USA

There are better places to watch birds, and much better places for wildlife in general, in the United States. But within its seventy-four square miles Aransas is the winter home for the fifty odd whooping cranes that spend the summer in Canada's Wood Buffalo National Park before flying the length of the continent to the Gulf Coast of Texas.

Aransas is a low-lying lagoon area intersected with creeks and marshes. There is nothing particular about it that picks it out from the surrounding land and yet year after year the whooping cranes return here and nowhere else on earth. In many parts of the world such a scarce and obvious animal would be locked away and few would be allowed to enjoy it—not so in the USA. Over 20,000 people annually call to see less than fifty birds—of course no one sees them all, they spend the winter scattered over the reserve in family parties, two birds here, three or four there. Large, specially fenced enclosures within which crane-preferred crops are grown attract a number of birds, as well as the whooping cranes.

But Aransas is not a one bird show. It has a list of over three hundred species to prove that. Most numerous and attractive are the other large waterbirds, including the once-rare, but now substantially recovered, roseate spoonbill. There are flocks of snow and lesser snow geese, though never in the numbers that occur at the specifically goose-based refuges, innumerable shore-birds and plenty of duck. Several hundred sandhill

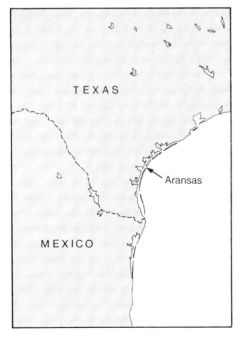

cranes frequent the refuge and there are many species of egrets and herons to be found, including the reddish egret.

Though not a major attraction, there are a few quadrupeds in the area including alligator, white-tailed deer, armadillo, raccoon and bobcat, but they are not often seen. Wild turkeys are more frequently encountered and other non-aquatic birds include white-tailed hawk, caracara, pyrrhuloxia, Brewer's blackbird, seaside and savannah sparrows, vermilion flycatcher and possibly bald eagle.

Visiting: The Refuge is open to visitors daily on the road south of Austwell, which is the best place to stay. There are tracks open to the public and well-marked blinds, though certain areas are sacrosanct for the whooping cranes.

Species of particular interest

White-tailed Deer
  Odocoileus virginianus
Whooping Crane
  Grus americana
Sandhill Crane
  Grus canadensis
American Egret
  Casmerodius albus
Reddish Egret
  Dichromanassa rufescens
Louisiana Heron
  Hydranassa tricolor
Green Heron
  Butorides virescens
White Ibis
  Plegadis chihi
Wild Turkey
  Meleagris gallopavo
Black Skimmer
  Rynchops nigra
Fulvous Tree-duck
  Dendrocygna bicolor
Roseate Spoonbill
  Ajaia ajaja
Lesser Snow Goose
  Chen hyperborea
Snow Goose
  Chen caerulescens
White-tailed Hawk
  Buteo albicaudatus
Common Caracara
  Caracara cheriway
Seaside Sparrow
  Ammospiza maritima
Savannah Sparrow
  Passerculus sandwichensis
Pyrrhuloxia
  Pyrrhuloxia sinnata

Aransas is not solely for the whooping cranes.
Louisiana herons (left) stalk the tangled
backwaters, while snowy egrets (right) perch,
yellow-footed, in the treetops. A host
of shorebirds use the Refuge as a
migration staging post, including (above)
pectoral sandpipers in the spring.

Arnhemland, Australia

Several thousand square miles of some of the most primitive land left on earth—that's Arnhemland in Australia's Northern Territory. 'Top End' they call it out there—top-end of the Territory. Roads are few and generally unmetalled. What's more they don't go anywhere either, or at least they don't go anywhere unless you're an aborigine. At the end of the road from Darwin lies the Aboriginal Reserve—entry prohibited. This is four-wheel drive territory, but then so too are many of our paradises.

Leaving Darwin, the obvious base, eastwards the road is metalled and progress is fast to the CSIRO village.

Fogg Dam is a waterbird haven and egrets and herons vie for attention. Pelicans, dotterels, kites, storks, the Australian crane the brolga, duck, all can be found at the lake and the sandy wallabies often reach plague proportions.

Continue on through the Marrakai Plains. Only twenty-five miles of good road left now, and wild water buffalo are numerous. From here the country is thickly forested, on to Cooinda the heart of western Arnhemland, and quaintly known as 'Jim Jim'. This is the end of the line for the average tourist, but it is also an excellent centre with innumerable swamps, lagoons and vast numbers of

A land of swamps and forests, Arnhemland is a home to a wide variety of animals. Gaily coloured rainbow bee-eaters (above left) flit overhead and dive into their unusual burrows, excavated in level ground. Reptiles of staggering variety include the frilled dragon (below left), which raises its ruff of skin to make it appear more ferocious than it is, while the marshy splashes support huge populations of the black and white magpie geese (below).

birds. Rainbow bee-eaters swoop low over the ground and cockatoos and galahs abound. There are parrots and honey-eaters, doves and even the great bowerbird. Goose Camp, near Nourlangie, is a great spot for aquatic species and the concentrations of magpie geese may reach 100,000 in the area. Unfortunately many of the best spots, including this, are out of bounds to the ordinary tourist, though permission can be granted by the ranger at Nourlangie. There are thousands of Australian jabirus (a kind of stork, not to be confused by the globe-trotting naturalist with the South American jabiru which is a completely different species) and tens of thousands of whistling duck and egrets. Then there are the white ibis, the whistling kites, the brolgas and pelicans, the black-necked stilts and so on. But this is not all in this fascinating and rich landscape. The Woolwanga Reserve is a beautiful area of savannah forest, monsoon scrub and shallow lagoons. Here there are honeyeaters, cockatoos, kingfishers and others.

A permit is required to visit Nourlangie Rock with its aboriginal rock paintings, but it is worth it for the paintings let alone the chestnut-quilled rock pigeon, black-banded pigeon and white-lined honeyeater, all of which are to be found only in the western part of Arnhemland. There are also rock wallabies, the rock-haunting ringtail possum and various small marsupials. Most of the smaller animals are nocturnal and are merely glimpsed as they flit across the road in car headlights. Native cats, however, will stop in the road when blinded by headlights.

The rivers boast the large estuarine crocodiles which may grow to over twenty-five feet. Protected only in 1970, these reptiles were almost exterminated by hunters. The smaller freshwater Johnstone crocodile is far more numerous and easy to see and there are other interesting reptiles including the frilled lizard and some of the so-called dragons.

Visiting: Start out in four-wheel drive from Darwin. The Arnhemland Highway provides surfaced motoring to South Alligator River. Access to Cooinda (tourist camp and petrol station) simple, but permits from the National Parks adminstration in Darwin or permission from the ranger at Nourlangie may be necessary thereafter.

Australian cranes, or brolgas, breed in Arnhemland. Their elaborate courtship dances are a magnificent spectacle and are the basis of many Australian aboriginal tribal dances.

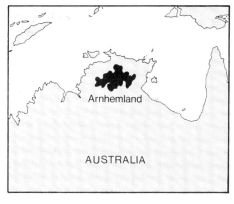

Arnhemland

AUSTRALIA

Species of particular interest

Sand Wallabies
  Protemnodon spp.
Rock Wallabies
  Petrogale spp.
Northern Black Wallaroo
  Macropus robustus
Rock-hunting Ringtail Possum
  Petropseudes dahli
Harvey's Marsupial Mouse
  Antechinus bilarni
Native Cat
  Satanellus hallucatus
Sugar Glider
  Petaurus breviceps
Estuarine Crocodile
  Crocodylus porosus
Freshwater Crocodile
  Crocodylus johnstoni
Frilled Lizard
  Chlamydosaurus kingii
Magpie Goose
  Anseranas semipalmata
Jabiru
  Xenorhynchus asiaticus
White-breasted Sea Eagle
  Haliaetus leucogaster
Whistling Kite
  Haliastur sphenurus
Pied Heron
  Ardea picata
White Ibis
  Threskiornis mulucca
Straw-necked Ibis
  Threskiornis spinicollis
Brolga
  Grus antigone
Black-breasted Pitta
  Pitta erythrogaster
Great Bowerbird
  Chlamydera nuchalis

Atherton Tablelands, Australia

Set in the far north of Queensland in the tropics, Atherton Tablelands is a superb, high altitude region of beautiful scenery and immense attraction to tourism. The lake created by the Tinaroo Dam is surrounded by hilly forests and visited by over one hundred thousand tourists a year.

Basically volcanic in origin, the Tablelands cover over 3000 square miles of hills, forests and lakes at an average height of 2500 feet. Along with the natural wonders of Millaa Millaa Falls, the natural rain forest along the Kennedy Highway, the curtain fig trees at Yungaburra and elsewhere, the Chillagoe Caves and Millstream Falls, there are golf courses, bowling greens, motels, camp sites and all the other accoutrements of the contemporary approach to the 'outdoors'.

Strange as it may seem wildlife has survived the onslaught of modern leisure. At one time the Tablelands were covered with rain forest; now only a few pockets of this fascinating habitat, so rich in wildlife, remain. What was once rain forest has been cleared for the cultivation of sugar-cane and tobacco. The remnants, together with the few undrained swamps and open eucalypt forests, provide the major habitats for the birds and native mammals of the area. A few species have adapted to cultivation but the larger mammals are found only in natural habitats.

The natural forests are true jungle. Vines grow luxuriously and figs include the strangler type that takes root in other trees and slowly surrounds and engulfs its host until it dies. Dead and dying trees are covered with epiphytic orchids and ferns, all giving an impression of jungle and making travel extremely difficult.

Among the most interesting animals are musk and rufous rat kangaroos and various other marsupials. The pademelon wallaby and black-tailed wallaby can both be found as can Lumholtz's tree kangaroo. The echidna, or spiny anteater, searches the forest floor and the duck-billed platypus lives among the jungle swamps and pools. The koala, scarce throughout its range, can still be found in numbers among the eucalypt stands and the various possums and native rats and mice find a true haven in the Tablelands.

But while large mammals do find sanctuary in the remaining areas of rain forest, the birds of this habitat in the north-eastern corner of Australia are truly exciting. In particular that extraordinary flightless, forest dweller, the cassowary, finds conditions here absolutely ideal. With its horny brow breaking its path it

can run through the jungle at astonishing speeds. But there are bowerbirds here too. Strange, ill-understood birds that create huge and elaborate bowers to entice females and which serve no other purpose than courtship. The magnificent satin bowerbird is one of the most beautiful birds in the world—with or without a bower. A vast selection of pigeons finds the natural habitat to its liking and there are flycatchers, pittas, logrunners, Australian robins, and, of course, the megapodes—brush turkeys with a breeding biology quite unique in the animal world. Like the other Australian megapodes, brush turkeys create a huge mound within which to lay their eggs. Incubation is then performed by the heat of the sun aided by the male, who removes or adds extra earth to maintain an equable temperature. Remarkably the chicks burrow their way to the surface on hatching and run away to fend for themselves without parental aid.

But while the natural areas attract perhaps the most interesting species, the cleared and cultivated areas boast many larger and perhaps more spectacular birds. The Australian crane, the brolga, leaps about the swamps in its elaborate courtship dances and the sarus cranes are nearby to make for confusion—both are grey with maroon head patterns. The jabiru stork scavenges over the marshes and black and white magpie geese gather in feeding groups, looking strangely primeval and ungainly as they fly. The uncanny noise of the kookaburra too seems primitive and there are a host of honeyeaters and woodswallows. Pygmy geese, of the Australian variety, whistling tree-ducks, jacanas, rollers, bee-eaters all add to the wealth of species to be found in this quite splendid area.

Of this vast region several small areas stand out as worthy of exploration. Lake Barrine, a clear crater lake, has much original rain forest and can produce the amethystine python. Lake Eacham is good for cassowary and tooth-billed bowerbird, and scrub turkeys are attracted to picnic grounds. Fresh-water tortoises feed from the pier.

Visiting: Free access by road almost everywhere. Vast range of accommodation available.

Species of particular interest

Musk Rat Kangaroo
  Hypsiprymnodon moschatus
Rufous Rat Kangaroo
  Aepyprymnus rufescens
Pademelon Wallaby
  Thylogale spp.
Black-tailed Wallaby
  Protemnodon bicolor
Sand Wallaby
  Protemnodon agilis
Lumholtz's Tree Kangaroo
  Dendrolagus lumholtzi
Duck-billed Platypus
  Ornithorhynchus anatinus
Koala
  Phascolarctos cinereus
Brolga
  Grus rubicunda
Jabiru
  Xenorhynchus asiaticus
Magpie Goose
  Anseranas semipalmata
Cassowary
  Casuarius casuarius
Brush Turkey
  Alectura lathami
Yellow-breasted Sunbird
  Nectarinia jugularis
Golden Bowerbird
  Sericulus aureus
Satin Bowerbird
  Ptilonorhynchus violaceus
Queen Victoria Riflebird
  Ptiloris victoriae

High among the forests of the Atherton
Tablelands the beautiful satin bowerbird (left)
constructs its courtship bower of twigs to
attract a mate. Frequently decorated with
colourful objects, bowers are kept religiously
tidy by the birds.

   Not all kangaroos are hopalongs.
Lumholtz's tree kangaroo (below) spends its
life among the treetops—a niche that in
other continents is occupied by monkeys.

Barro Colorado, Panama Canal Zone

With the cutting of the Panama Canal the Chagres River valley was flooded and many of what had formerly been hill tops became islands. Following the enthusiasm of Thomas Barbour a Reserve-Laboratory was established on one of the islands in 1923 and on his death its administration was continued by the Smithsonian Institute. Covering a mere 3500 acres Barro Colorado, as it is generally known, is a self-contained and completely protected unit that enjoys the same balance and variety of animals previously found in the surrounding country. Canal construction raised the water level by some eighty feet and covered 165 square miles, but it did so so gradually that the animals could adjust rather than become highly concentrated.

Barro Colorado is under three miles wide and is situated in Gatun Lake mid-way between the Atlantic and the Pacific. It is mainly forested, and boasts a complete cross-section of the Central American fauna. There are pumas, jaguars and ocelots, peccaries, sloths, agoutis, coatis, monkeys and over 200 species of birds—all on a tiny island that can be completely explored and where the fauna has never been persecuted. Howler monkeys are particularly obvious in the evenings, calling from the trees where their prehensile tails help to support them. On the ground the coatis, intensively studied here, search the leaf litter or even the research buildings or living quarters for their food. And where else is one likely to get even the chance of seeing a jaguar? Then there are the birds, among the most exotic and rich in the world. Manakins, many performing their strange communal displays, tanagers, motmots, puffbirds, toucans and parrots—species that will take weeks to see and identify. Iguanas, huge reptiles, climb the trees out of the water in which they spend so much of their time.

Only a short distance away ply the world's great merchant ships, unaware of the miniature wildlife paradise of Barro Colorado.

Visiting: The island is not a park or a tourist centre, it is a research centre. Scientists may visit it with permission from the Smithsonian Institute.

Species of particular interest

Jaguar
  Panthera orca
Puma
  Felis concolor
Ocelot
  Felis pardalis
Collared Peccary
  Tayassu tajacu
White-lipped Peccary
  Tayassu albirostris
Coati
  Nasua nasua
Tapir
  Tapirus terrestris
Sloth
  Bradypus infuscatus
Agouti
  Dasyprocta spp.
Tayra
  Eira barbara
Parrots
  Psittacidae
Hummingbirds
  Trochilidae
Gould's Manakin
  Manacus vitellinus
Oropendolas
  Gymnostinops spp.
Tanagers
  Thraupidae
Motmots
  Motmotidae
Puffbirds
  Bucconidae
Toucans
  Ramphastidae
Flycatchers
  Tyrannidae

Many otherwise rare Central American animals can be seen to advantage on the carefully protected island of Barro Colorado. Coatimundis (right) frequently haunt the buildings where scientists work, while the elusive jaguar (below right) may be seen at a marshy pool when it comes to drink.

Point Barrow, Alaska, USA

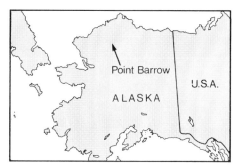

By their very nature wildlife paradises are often difficult of access, remote and inhospitable. Very few areas of the world, where wildlife is a major attraction, can be approached as easily as the East African game parks. But while others are also difficult to reach, few can be so ruggedly austere as Alaska.

The 49th State of the Union is said to be on the verge of another economic boom— it endured the Yukon gold rush for the few brief years of its existence. Now oil and other mineral wealth is about to transform a state that already enjoys the reputation of having one of the most expensive price systems on earth. Today it is not the Klondike but the North Slope that is booming.

The North Slope is the gently sloping, often flat landscape that runs northwards from the Brooks Range of mountains. It's a vast area frozen for most of the year but underlain with some of the world's richest oil-bearing strata. Point Barrow is an Eskimo village that is fast becoming an international centre of the high-powered oil business. By sheer good fortune this means that naturalists can visit the area more easily—a wilderness that boasts centrally heated hotels, baths etc.

Though it is barren (or comparatively so) in winter, the late spring sees the thaw that brings birds by the million to the North Slope. Duck and geese, gulls and skuas, sandpipers and plovers are everywhere among the pools and mosses that overlay the permafrost a few feet below.

But it is not only the birds that migrate through Alaska. About half a million caribou leave the woodlands to the south to come up here every spring. Apart from the great migrations of plains game in Tanzania there is nothing else like this— vast herds of large animals on the move. Many traverse the Anaktuvuk Pass where the Nunamlut Eskimoes lie in wait for them. Wolves too wait in ambush for the herds but their numbers are declining and their unpopularity will probably eliminate them altogether before long.

In the barren lands of Point Barrow,
long-tailed skuas fly overhead, ever watchful
for an unguarded nest. Pirates at sea, the
birds find eggs and small mammals a fine
source of food on their tundra
breeding grounds.

The barren-ground grizzly bear is another progressively scarce inhabitant of the North Slope but it is omnivorous and can manage on a diet of roots and ground squirrels.

Musk ox are nowhere plentiful but can be found here and there in small herds. The strangely bedecked antlers of the moose are more frequently seen along the rivers with their growth of stunted trees, while away to the south, among the Brooks peaks, the all white dall sheep live on the most inaccessible peaks. Fortunately they gather in numbers in places like Atigun Canyon where nearly a thousand can be found in May.

Though these large animals are a great attraction, the average visitor will be thrilled by the birds that he finds within yards of his hotel. It is the tameness of many species that is so entrancing. Lapland longspurs (buntings in Britain) perch on the rooftops of the small settlements; sandpipers have to be almost lifted from their eggs; snowy owls ghost round the intruder staring arrogantly and perhaps swooping menacingly to scare him off. In the twilight of the midnight sun, loons call bewitchingly over the marshes, a fabulous sound.

Visiting: Air transport is one of Alaska's bargains—everything else is so expensive. There are hotels at Point Barrow, but for real exploration camping is essential.

Species of particular interest

Barren-ground Caribou
  Rangifer tarandus
Moose
  Alces alces
Grizzly Bear
  Ursus americanus
Wolverine
  Gulo gulo
Wolf
  Canis lupus
Dall Sheep
  Ovis dalli
Musk Ox
  Ovibus moschatus
Brown Lemming
  Lemmus spp.
Collared Lemming
  Dicrostonyx spp.
Yellow-billed Loon
  Gavia adamsii
Long-tailed Duck
  Clangula hyemalis
King Eider
  Somateria spectabilis
Snow Goose
  Chen caerulescens
White-fronted Goose
  Anser albifrons
Grey Phalarope
  Phalaropus fulicarius
Pectoral Sandpiper
  Calidris melantos
Baird's Sandpiper
  Calidris bairdii
Semi-palmated Sandpiper
  Calidris pusillus
Long-tailed Skua
  Stercorarius longicaudus
Snowy Owl
  Nyctea scandiaca
Lapland Longspur
  Calcarius lapponicus

Grizzly bears along the streams and caribou on the fields are two of the attractions of the Barrow area of northern Alaska. During the summer run of sockeye salmon the bears feast on the spent fish (right), while the caribou graze on the mosses of the tundra (above) before migrating southwards to the shelter of the great coniferous forests.

Blue Mountains, Australia

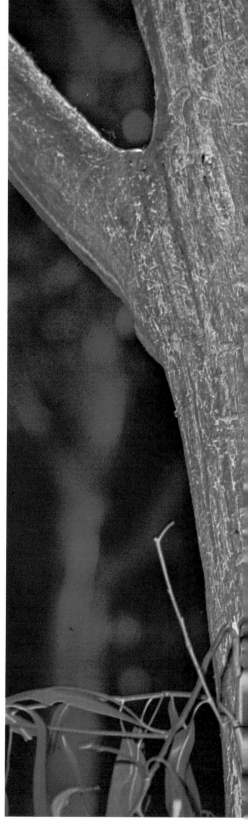

Sixty miles west of Sydney lie the Blue Mountains. Of no great altitude their sheer sandstone cliffs, standing boldly from the sea of their forested lower slopes, effectively barred the expansion of the convict colony until three explorers found a way through in 1813. Their route is now part of Blue Mountains National Park.

Deep canyons, wooded slopes, wet jungles, open plains, orchards, gentle heathland plus waterfalls and many caves: that's the Blue Mountains. But they are also a holiday playground with easy access by car and electric train. There are cable cars and picnic sites, hotels and motels, golf courses and pony treks and plenty of wildlife at its best.

Even the endemic mammals perform well here and the rock wallabies are particularly numerous and attractive. The large gray kangaroo often comes out into the open, and black-tail and red-necked wallabies are numerous. The stands of eucalyptus are still frequented by that charmer the koala, here, as in most cases, unfortunately becoming quite scarce. Most of the other mammals are nocturnal but it's worth searching around with a flashlight at night.

The duck-billed platypus can be found

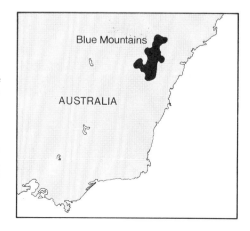

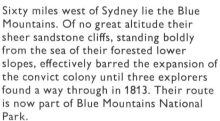

Koalas (below), exterminated over large areas of their range, are still plentiful among the eucalyptus forests of the Blue Mountains. Harmless and cuddly, they are a constant attraction to tourists and wildlife enthusiasts. The marsupials of Australasia have, in the absence of competitors, evolved to fill a variety of ecological niches. The brush-tailed rock wallaby is just one that has found a hyrax-like life style among the rocky ground suitable to its needs (left).

Common and widespread, the colourful galahs (left) may be seen at their nest holes or gathered in large numbers round a favoured waterhole. A boobook owl (right) flies into its nest in a broken stump with a small snake for its youngsters.

in many of the clear creeks, particularly in the remoter valleys. And the other egg-laying mammal the echidna, or spiny anteater, can be found at the wildlife sanctuary at the Jenolan Caves. There are plenty of other birds and animals here. Few visitors can fail to be impressed by the Grand Arch which forms the entrance to one of the most impressive of cave systems. Huge arrays of stalactites and stalagmites, plus a host of other formations, attract thousands of sightseers every year. But not all of them see the brush-tailed rock wallabies that descend to the cave area every night and which can be persuaded by bread to venture within a few feet. When the sun goes down there's a good chance of a brush-tailed possum coming down a tree or of a wombat crossing the road.

Then there are the birds, a couple of hundred species of them varying from the superb lyrebird in the fern gullies to the rock warbler which is only found in sandstone country in the vicinity of Sydney. There are wedge-tailed eagles, parrots and their allies, the boobook owl, tawny frogmouth and so on. Go in spring and the area is alive with wild flowers.

Visiting: The Park is easily reached from Sydney and there is plenty of accommodation. The Jenolan Cave Hotel is run by the New South Wales Tourist Department and is an attractive spot for those who wish to stay overnight and see mammals.

Species of particular interest

Koala
   Phascolarctos cinereus
Brush-tailed Rock Wallaby
   Petrogale penicillata
Grey Kangaroo
   Macropus canguru
Black-tailed Wallaby
   Protemnodon bicolor
Red-necked Wallaby
   Protemnodon rufogrisea
Brush-tailed Possum
   Trichosurus vulpecula
Ring-tailed Possum
   Pseudocheirops archeri
Sugar Glider
   Petaurus breviceps
Greater Glider
   Petaurus norfolcensis
Wallaroo
   Macropus robustus
Wombat
   Vombatus ursinus
Echidna
   Tachyglossus aculeatus
Duck-billed Platypus
   Ornithorhynchus anatinus
Wedge-tailed Eagle
   Aquila audax
Superb Lyrebird
   Menura novaehollandiae
Tawny Frogmouth
   Podargus strigoides
Eastern Rosella
   Platycercus eximius
Crimson Rosella
   Platycercus elegans
Rainbow Lorikeet
   Trichoglossus haematodus
Kookaburra
   Dacelo gigas
Mistletoebird
   Dicaeum hirundinaceum
Rufous Fantail
   Rhipidura rufidorsa
Wonga Pigeon
   Leucosarcia melanoleuca
Boobook Owl
   Ninox novaeseelandiae
Pied Currawong
   Strepera graculina
Satin Bowerbird
   Ptilonorhynchus violaceus

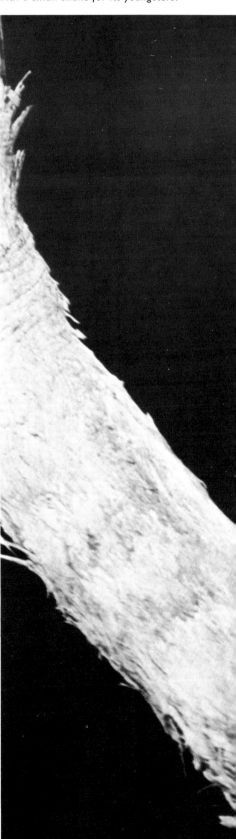

Camargue, France

If anywhere in Europe can properly be called a wildlife paradise, it must be the delta of the River Rhône. From its source, high up in the Swiss Alps, this great river passes through France, separating the French Alps from the Massif Central. From Lyon it flows directly southward to Avignon and Arles before dividing up into three separate channels at its Mediterranean delta. Here the Romans enjoyed the lush sub-tropical climate of the Camargue, leaving amphitheatres and monuments behind in the charming city of Arles.

The Camargue is a wilderness. Huge shallow reed-fringed lagoons cover the landscape interrupted only by the occasional farmsteads surrounded by tall protecting trees. Protection, that is, against the dreaded 'mistral'—a wind that starts in the Alps and is funnelled down to reach gale force as it spends itself over the sea. It is an unpredictable foe, and though it only lasts a few days, tempers can fray while it blows.

Water is everywhere, though large areas of the Camargue have been transformed into salt pans in the south, and rice fields in the north. A network of roads cuts here and there across the delta, which is dominated by the huge Etang de Vaccares. Here, egrets fish among the shallows while rafts of duck, including ferruginous, and red-crested pochard sit out the splashy waves. Black kites comb the shoreline out-numbered only by the marsh harriers that quarter the reed beds. Short-toed eagles fly overhead, while avocets, black-winged stilts, pratincoles and Kentish plovers find nest sites on the adjacent marshes as they dry in the summer sun. Small birds there are in profusion. Moustached and Savi's warblers, penduline tits, and hosts of marsh terns, including numerous whiskered terns, swoop over the rice paddies. Rollers flit along the telegraph wires, while bee-eaters gather in hundreds at their colonies.

Eastwards across the main arm of the Rhône lie the stony wastes of La Crau, the delta of the young, virile river Durance. Large pebbles stretch as far as the eye can see, broken only by the occasional sheep pen and the piles of stones laboriously collected by some poor shepherd. Black-eared wheatears, interesting larks, pin-tailed sandgrouse and lesser kestrels are the attraction. But at Enteressen, the local rubbish tip is alive with carrion eaters including Egyptian and griffon vultures, that have come down from the hills.

Vaccares is the heart of the Camargue,

Flamingoes feed among the great lagoons of the Camargue while the heat reduces the background to a shimmering haze.
Several species, like these greater flamingoes, find in the great delta of the Rhône their most northern European outpost.

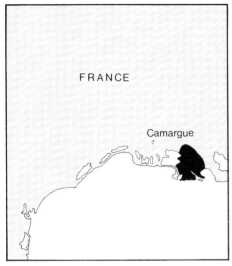

Species of particular interest

Wild Boar
  Sus scrofa
Flamingo
  Phoenicopterus ruber
Marsh Harrier
  Circus aeruginosus
Montagu's Harrier
  Circus pygargus
Black Kite
  Milvus migrans
Short-toed Eagle
  Circaetus gallicus
Egyptian Vulture
  Neophron percnopterus
Red-crested Pochard
  Netta rufina
Kentish Plover
  Charadrius alexandrinus
Wood Sandpiper
  Tringa glareola
Greenshank
  Tringa nebularia
Avocet
  Recurvirostra avosetta
Black-winged Stilt
  Himantopus himantopus
Black Tern
  Chlidonias niger
Whiskered Tern
  Chlidonias hybrida
Caspian Tern
  Hydroprogne tschegrava
Mediterranean Gull
  Larus melanocephalus
Slender-billed Gull
  Larus genei
Pin-tailed Sandgrouse
  Pterocles alchata
Roller
  Coracias garrulus
Bee-eater
  Merops apiaster
Cetti's Warbler
  Cettia cetti
Great Reed Warbler
  Acrocephalus arundinaceus
Fan-tailed Warbler
  Cisticola juncidis

and though birds as exciting as slender-billed gulls can be found among the gull and tern colonies, most visitors will want to see the flamingoes. Over five thousand nest irregularly on the lagoons to the south, but many can be seen far out in the shallows of this vast lagoon—a pink shimmer in the heat haze of the middle of the day. Flamingoes nest only at one other site in Europe.

Visiting: The best way to see the Camargue is by horseback. Visitors can stay at Arles, or at the houses of the guardians. Write to Zoologiques et Botaniques de Camargue,
2 rue Honoré Nicholas, Arles,
B. du R., France. You can camp at various places. Do not miss the gypsy festivals at Arles and Stes Maries de la Mer.

The long-legged waders find the varying depths of water of the Camargue ideally suited to their feeding methods. The saline lagoons near the coast attract both avocets (right) and black-winged stilts (below) to breed in large numbers, though most other waders are simply birds of passage.

Carlsbad Caverns, New Mexico, USA

It's strange to think of a wildlife paradise underground but that really is the centrepiece of the Carlsbad area—a purely underground wildlife spectacle. Carlsbad Caverns lie in New Mexico 150 miles from El Paso and form the largest system of caves in the world. As yet they are not completely explored, but the average visitor takes a three mile trip while expert pot-holers have penetrated over 1000 feet below the surface.

Being the largest in the world is an attribute in any superlative-conscious country and the caves are visited by thousands of tourists every year. There is a car park for 600 cars, an elevator to penetrate deep into the cave system and all the usual paraphernalia of a tourist attraction—bubble gum but no candy floss. Yet in the midst of all this is one of the greatest wildlife spectacles.

No one has counted the bats of Carlsbad Caverns; people just say millions, and every summer evening these millions leave the caves to seek food outside. It's an impressive sight—a great sight. The roosting caves inside the system are not usually open to the public, though *bona fide* scientists may obtain permission. But the sight of all these millions of animals clinging to the ceilings of the caves is, if not unique, at least spell-binding. That the millions of bats consist of no less than eleven species usually escapes most visitors who simply thrill at the sight. But there are Mexican free-tailed, by far the most numerous of the species present, with fringed myotis, western pipistrelle, pallid and lump-nosed bats.

Visitors awaiting the evening flight can while away the time exploring the rest of the seventy-seven square miles that make up Carlsbad Caverns National Park. This is mainly desert country with cacti (bloom in May and June) and other flowers changing the face of the arid landscape. Mule deer and pronghorn are the largest of the mammals but there are jack-rabbits, skunks, raccoons, desert foxes and an interesting array of birds. Of course, no ornithologist would come here just to see birds but he'd be pleased to spot canyon and rock wrens at the entrance to the caves, and pyrrhuloxias, brown towhees and rufous-crowned sparrows as a bonus.

Visiting: The caverns are 27 miles from Carlsbad which offers the nearest accommodation. An entrance fee is payable. True 'batmen' should contact Box 111, Carlsbad, New Mexico 88220.

Species of particular interest

Pronghorn
  Antilocapra americana
Cottontail
  Sylvilagus spp.
Skunk
  Mephitis mephitis
Raccoon
  Procyon lotor
Mexican Free-tailed Bat
  Tadarida spp.
Fringed Myotis
  Myotis sp.
Western Pipistrelle
  Pipistrellus spp.
Pallid Bat
  Antrozous spp.
Turkey Vulture
  Cathartes aura
Cactus Wren
  Campylorhynchos brunneicapillum
Rock Wren
  Salpinctes obsoletus
Whip-poorwill
  Caprimulgus vociferus
Road Runner
  Geococcyx californianus
Brown Towhee
  Pipilo fuscus
Rufous-crowned Sparrow
  Aimophila ruficeps
Blue Grosbeak
  Guiraca caerulea
Pyrrhuloxia
  Pyrrhuloxia sinuata

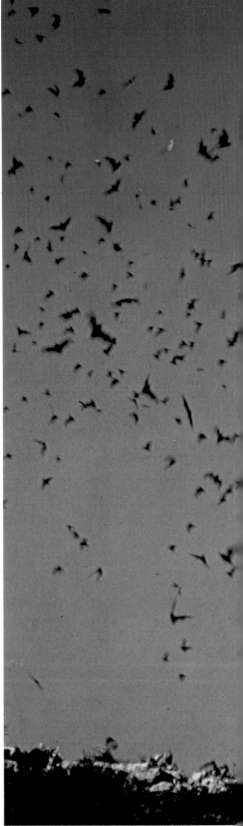

Caroni Swamp and Springhill, Trinidad and Tobago

The island of Trinidad, together with its satellite Tobago, lies just off the coast of South America. Adjacent Venezuela, together with neighbouring Colombia, is the richest area in the world for birds and it is not surprising that these offshore islands should share in this wealth. Only the shallow Gulf of Paria separates Trinidad from Venezuela and the narrow channel, the Boca Grande, is but a few miles across. Yet this short distance does make a difference to the fauna and flora of Trinidad. The yellow-headed caracara, for example, manages to cross the channel but only as far as the small island of Chacachacare. It ventures no further. As if to compensate, other, more marine, species are found on Trinidad and Tobago, but are not found on the mainland.

Trinidad covers about 1800 square miles, while Tobago is a little over 100 square miles. The larger island has mountains covered with dense tropical forests and with cocoa plantations in the valleys. There are areas of savannah together with some of the most exciting swamp areas in the world. Foremost among these are the Caroni Swamps on the west coast, south of Port of Spain. Here a great collection of waterbirds congregate, among which scarlet ibis and roseate spoonbill are the top prizes.

The ibis flight in to roost against the setting sun—an unforgettable spectacle. Among the herons to be found at Caroni are little blue, snowy egret, that trans-Atlantic colonist the cattle egret, the rare red-necked heron, black-crowned night heron, tiger-bittern, etc. The strangely-endowed boat-billed heron nests in the swamps and the anhinga can be seen from time to time. Naturally there are many smaller birds, several of which are extremely colourful.

But for real colour it is necessary to penetrate the forests (jungles would be a better term) of the interior and the hills. Though many tracks and roads provide an excellent method of exploration, the visitor who did not visit the Springhill Estate of the Asa Wright Nature Center would have missed a significant experience. There a hotel-guest-house has been established to cater for visiting naturalists intent on birds, and there are a lot of them. Jacamars, toucans, manakins (their displays have been better studied here than anywhere), cotingas, ovenbirds, antbirds, tinamous can all be found in profusion, and there are staff on hand to help with identification. There are hummingbirds busily feeding around the plantations, and hawks and kites overhead. The fortunate may see the swallow-tailed kite, while if circumstances permit, visits to the caves for oilbirds may be allowed. These spectacular birds emerge at night to feed on fruit and are almost unique in the bird world in using an echo-location system for flying in the total darkness of their cave refuges. Unfortunately disturbance by tourists such as ourselves is interfering with the survival of the species. Over 500 species of butterflies have been recorded and seventy mammals, including armadillo, peccary, ant bear, anteater and ocelot.

Visiting: International air services to Port-of-Spain. Paths across Caroni Swamp, and boats may be locally hired. Trips can be organized by Hub Travel and permits may be obtained from the Tourist Board—both in Port-of-Spain. The Asa Wright Centre, Port-of-Spain, also supply information on the Swamp.

Species of particular interest

Little Tinamou
Crypturellus soni
Scarlet Ibis
Eudocimus ruber
Roseate Spoonbill
Ajaia ajaja
Little Blue Heron
Florida caerulea
Snowy Egret
Egretta thula
Cattle Egret
Bubulcus ibis
Red-necked Heron
Hydranassa tricolor
Black-crowned Night Heron
Nycticorax nycticorax
Tiger-bittern
Tigrisoma lineatum
Boat-billed Heron
Cochlearius cochlearius
Anhinga
Anhinga anhinga
Swallow-tailed Kite
Elanoides forficatus
Blue-crowned Motmot
Motmotus momata
Channel-billed Toucan
Ramphastos vitellinus
Oilbird
Steatornis caripensis
Bearded Bellbird
Procnias averano
Tyrant Flycatchers
Tyrannidae
Swallow-tanager
Tersina viridis

Chinchas Islands, Peru

The Chinchas probably hold more birds than any other similarly sized area in the world. The eye of the visitor is inevitably drawn to the silent rows of brown pelicans (left) that stand sentinel along the rocks and the graceful Inca terns (top) that wheel overhead and nest on the precipitous ledges. The near-white feet of the blue-footed booby youngster (bottom) will later turn to its parents' turquoise colour.

The cold Humboldt Current sweeps along the coast of Peru creating some of the richest seas in the world. The feeding is so rich that fish virtually swarm along the barren coast, while overhead flies one of the greatest concentrations of seabirds on earth. Because the sea is cold there is little evaporation and thus little rainfall. The result on the land is aridity—parched earth and meagre crops. The result on the islands that lie offshore is the rich deposits of guano laid down by the millions of birds that breed on them— there is just no rain to wash them away.

When first discovered the Islas de Chincha, lying off the coast near the port of Pisco some two hundred miles south of Lima, were covered with guano to a depth of 130 feet. Within a few generations these had been cut back to the bed-rock by nitrate-hungry countries in need of a prop to their own agriculture. Between 1851 and 1872 ten million tons of guano were removed from the Chinchas alone. So serious were these depradations that a special administration was established to protect the islands from over-exploitation, and the birds from destruction and disturbance. This was nothing new, however; the Incas had a harsh system of punishments for similar offences.

To the wildlife enthusiast the Chinchas are a paradise· Up to twenty million birds are found along this coast, and these islands are the most important and densely populated. Most numerous of all are the guanay cormorants. Literally millions and millions of birds spread out over the entire surface of the islands, all nesting on the accumulated droppings of their ancestors. Peruvian boobies and brown pelicans are next in order of importance as guano producers, though there are also numbers of blue-footed boobies. The pelicans are perhaps the most spectacular of the birds. They dive from the air like darts, throwing their wings backwards at the last minute as they enter the sea. On the Chinchas the three major species can total up to four or five million birds.

The number of birds varies from year to year over about a seven year cycle. Some years, instead of a cold current sweeping along the coast, the world's marine equator moves southwards and warm water moves in. This not only contains less food but also brings rain to these arid regions. Such occasions are disasters and the birds die by the million—there is just nowhere else they can go.

In the 1960s industrialized fishing began to affect the vast shoals of sardines. The concept was simple—why wait for the birds to produce the nitrates when the fish themselves could be caught and processed into fish meal. The result was over-fishing and a serious decline in the number of birds.

Other species such as the Peruvian penguin and Peruvian diving petrel can also be found here, though neither is of great importance as a producer of guano. The delightful grey Inca tern, resplendent with two white 'horns', flies gracefully around the cliffs. Kelp and gray gulls can be seen over the tide-line joined by large numbers of migrant Franklin's gulls, skuas and common and arctic terns from North America. Pink-footed shearwaters and Peruvian storm petrels breed nearby.

On the adjacent parts of the mainland the Andean condor (with its 12ft wingspan), nowhere near as rare as its Californian equivalent, descends to sea level. Robert Cushman Murphy describes the stomach contents of one bird as 'large pebbles, fish remains, bodies of diving petrels, parts of a penguin, the hoof of a pig, fresh eggs of guano birds, small bits of kelp, and the femur, radius and cartilages of a fur seal'. The fur seal is, in fact, almost the only mammal of note to be found at the Chinchas.

Visiting: Boats can be arranged to tour round the islands from Pisco, but landing must be arranged with the Compania del Guano.

Species of particular interest

Peruvian Fur Seal
 Arctocephalus australis
Guanay Cormorant
 Phalacrocorax bougainvillii
Brown Pelican
 Pelecanus occidentalis
Peruvian Booby
 Sula variegata
Blue-footed Booby
 Sula nebouxii
Peruvian Penguin
 Spheniscus humboldti
Peruvian Diving Petrel
 Pelecanoides garnotii
Pink-footed Shearwater
 Puffinus creatopus
Peruvian Storm Petrel
 Oceanites gracilis
Andean Condor
 Vultur gryphus
Kelp Gull
 Larus dominicanus
Gray Gull
 Larus modestus
Franklin's Gull
 Larus pipixcan
Arctic Tern
 Sterna paradisaea
Common Tern
 Sterna hirundo
Inca Tern
 Larosterna inca

Blue-footed boobies are one of the larger guano birds of the Chinchas. Widespread in the tropics they reach their greatest density on the richest bird grounds on earth.

Chitawan, Nepal

Chitawan National Park lies at the confluence of the rivers Rapti and Rue in the Terai region of central southern Nepal. It is less than ten miles from the Indian border and covers some 240 square miles of hilly jungle country dominated by mixed hardwood forest with sal, together with open areas of marsh and elephant grass. Chitawan is the largest remnant of the jungle that once covered the whole of the low-lying Terai region. A massive and successful anti-malaria campaign in the 1950s made the area suitable for human settlement and much jungle has been cleared for pasture and farming. Invading cattle and gathering of firewood are the major threats though poaching is a continuing problem. Tracks are few and the park is best worked by elephant *via* Tiger Tops Jungle Lodge.

The Park is dominated by the great Indian one-horned rhinoceros of which some 200, out of a total world population of 1000, are found at Chitawan. It is one of the best places in the world for this endangered species which is usually seen without problem. Tigers number less than fifteen and, though buffalo is put out as bait nightly during the season, visitors are far from sure of seeing this species. Among a wealth of other species chital, hog deer and barking deer are reasonably common and gavial are found at the Rapti-Rue confluence. Some three hundred bird species include darter, black-necked stork, grey fishing eagle and four species of noisy but attractive parakeets. For European ornithologists a wealth of *Phylloscopus* warblers poses fascinating problems of identification. Butterflies are beautiful and plentiful, though very under-worked.

Visiting: Chitawan can be reached by road but it is rugged and suitable only for four-wheel drive vehicles and outside the monsoon season. Royal Nepalese Airlines fly Kathmandu–Meghauli daily and visitors are greeted by elephant jungle-taxis. Tiger Tops is an idyllic jungle lodge (excursions and meals included in daily price). Book everything including flight through their offices: PO Box 242, Kathmandu, Nepal, or 226 W. 47th St., New York, NY10036, USA.

Through the grey morning mist of Chitawan National Park (left) the high Himalayan peaks may be seen in the background. The park has some two hundred Indian one-horned rhinoceros (below) out of a total world population of only a thousand, and is the second most important area in the world for this animal.

The dense jungle of Chitawan is a home to numerous langur monkeys (left) which swing from the treetops and scurry over the rocky hillsides. Above, along the Rapti River, gavials, the fish-eating crocodiles, haul out to rest open-mouthed in the harsh sun.

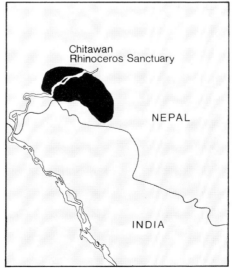

Species of particular interest

Indian Rhinoceros
  Rhinoceros unicornis
Tiger
  Panthera tigris
Leopard
  Panthera pardus
Wild Pig
  Sus scrofa
Sambar
  Cervus unicolor
Chital
  Axis axis
Hog Deer
  Axis porcinus
Barking Deer
  Muntiacus muntjak
Langur
  Presbytis entellus
Rhesus Monkey
  Macaca mulatta
Gavial
  Gavialis gangeticus
Mugger
  Crocodylus palustris
Indian Darter
  Anhinga rufa
Black Ibis
  Pseudibis papillosa
Open-billed Stork
  Anastomus oscitans
Black-necked Stork
  Xenorhynchus asiaticus
Gray Fishing Eagle
  Icthyophaga ichthyaetus
Egyptian Vulture
  Neophron percnopterus
Red Junglefowl
  Gallus gallus
Blue Peafowl
  Pavo cristatus
Pheasant-tailed Jacana
  Hydrophasianus chirurgus
Bronze-winged Jacana
  Metopidius indicus
Indian River Tern
  Sterna aurantia
Black-bellied Tern
  Sterna acuticauda
Large Green Parakeet
  Psittacula empatria
Rose-ringed Parakeet
  Psittacula krameri
Great Indian Hornbill
  Buceros bicornis
Crested Swift
  Hemiprocne longipennis
Ashy Swallow Shrike
  Artamus fuscus
Black-headed Oriole
  Oriolus xanthornus

Churchill, Canada

When the snow melts, the whole of Canada's northlands become a vast maze of lakes and swamps that teem with wildlife for the brief interlude before it all freezes up once more. During this short, hot summer a wildlife paradise on a gigantic scale stretches from the Bering Sea to Hudson Bay, an area that of its very nature is impenetrable except at a few isolated spots. Churchill is one of these.

Situated on the western shore of Hudson Bay at the mouth of the Churchill River, the surrounding area is a vast wilderness occupied because of its historical establishment as a trading post. In summer the colourful lichens and masses of arctic flowers bring the landscape to life. It's a rich landscape, where two great biomes, the tundra and taiga (the swampy northern coniferous forests) meet. To the north only the dwarf birch and willow can survive; to the south the forests stretch for thousands of miles.

Caribou and moose can be seen along with the occasional wolf. The more regularly-seen arctic fox finds the easy pickings provided by young birds a pleasant change from the hard, lean search of winter. As the thaw of spring frees the rivers, beluga whales move in as they have done for generations. While the autumn freeze-up gets under way migration brings an influx of polar bears to raid the dustbins and garbage dumps. They always come for the seal hunting, but scavenging is easier. Up to 200 may be found in and around the town but scientists are trapping them and flying them away to distant parts of the territory where they are less of a nuisance.

But it is the birds that provide the main attraction at Churchill. In spring in particular, huge numbers of migrants pour in spending time off feeding and awaiting the thaw of their breeding grounds to the north. For the snow geese of the central flyway, Churchill is the end of the line. Vast numbers breed a distance away to the south. Hordes of other wildfowl find that the pools and streams of the tundra provide a safe haven and a plentiful food supply and settle to breed. To spot over a hundred species in a couple of weeks in June is not difficult, and most species are very numerous. Hudsonian godwit, whimbrel, golden plover, arctic jaeger (skua), Lapland and Smith's longspurs (buntings) are all easy to find breeding locally.

To the south among the trees are gray jay and myrtle and blackpoll warblers. Rock ptarmigans are everywhere and

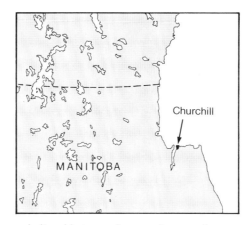

unbelievably tame. A peregrine usually uses the grain elevators of the town as a vantage point. And for the really dedicated, Thayer's gull, formerly considered a sub-species of the herring gull, mixes freely with that species in the town centre.

Visiting: Fly or take the train (two days) from Winnipeg. Book accommodation well in advance and take plenty of warm clothing.

Species of particular interest

Polar Bear
  Thalarctos maritimus
Beluga Whale
  Delphinapterus leucas
Wolf
  Canis lupus
Arctic Fox
  Alopex lagopus
Yellow-billed Loon
  Gavia adamsii
Canada Goose
  Branta canadensis
Rock Ptarmigan
  Lagopus mutus
Sandhill Crane
  Grus canadensis
Old Squaw
  Clangula hyemalis
Ruddy Turnstone
  Arenaria interpres
Hudsonian Godwit
  Limosa haemastica
Semi-palmated Sandpiper
  Calidris pusillus
Whimbrel
  Numenius phacopus
Arctic Skua
  Stercorarius skua
Smith's Longspur
  Calcarius pictus

In the bird-rich area of Churchill on Hudson Bay, common loons (top) wail out their mournful cries and nest along the edges of the tundra pools. Among the scattered vegetation willow ptarmigans, resplendent in a mixture of summer and winter plumage, may be seen (bottom).

Corbett National Park, India

Named after the famous tiger hunter and wildlife enthusiast Jim Corbett, this Park lies in the Himalayan foothills west of Nepal and to the north-east of Delhi. It was India's first national park and was first named Hailey National Park after the first governor of Uttar Pradesh, Sir Malcolm Hailey, in 1935. This area of the country has always been noted for its tigers and Jim Corbett was particularly adept at finding man-eaters. Now the Park, which covers 125 square miles, is one of the last strongholds of this much endangered animal.

The Ramganga River cuts deep into the Park, its ravines alive with orchids and other tropical plants. The surrounding hills are covered with a thick jungle of sal, silk-cotton and the light green shisham trees knotted together with creepers. The lowland plains are grassy and support a rich mammal fauna that is the major attraction to visitors and tigers alike— although for different reasons.

In the west of the Park the river has been dammed and a huge lake created right in the heart of the lowland area. This has proved of dubious benefit, for though water is usually an enhancing factor to wildlife areas, this particular reservoir has covered much valuable grazing in a park that is essential to the survival of the tiger and its prey. The construction work itself was of course extremely damaging to the park, its fauna and flora. It is to be regretted that even so small and famous a national park as Corbett is not immune from economic development.

The river and its pools are the home of mahseer (often called 'Indian salmon') and 'Indian trout'. Both provide food in abundance for the goonch (which is sometimes known as the 'freshwater shark'). All are taken by the long-snouted gavial, the fish-eating crocodiles that lie along the river banks all day long with their mouths wide open. Their relatives, the marsh muggers, are also common among the swampy backwaters. Like all crocodiles these are much sought after for their skins.

Where the Ramganga broadens out it runs across the flat open grassy plains beloved of deer. Here the elegant spotted deer or chital are quite common and hog deer are found. There is even a small herd of swamp deer or barasingha, though the largest herd of this rare deer is found to the east at Sukla Phanta in western Nepal. The edges of the plains are the natural haunt of tigers though, as elsewhere, these magnificent beasts are seldom seen

except at a kill. Elephant expeditions do sometimes manage to 'round up' a tiger but the number of elephants required is usually prohibitively expensive. The Park also supports small herds of wild elephant. Near Boksar is a place called Sagar Tal where the gods descend every January and which draws several hundred pilgrims each year. The day is called Makar Sankranti and is well worth noting and avoiding if the aim is to see wildlife.

The forests, which are dominated by sal, are the home of many birds and both rhesus and the larger grey langur swing among the tree tops. Parrakeets are particularly obvious and many Himalayan birds descend to winter and roam nomadically through the forests like tit flocks in a northern woodland.

Species of particular interest

Tiger
  Panthera tigris
Indian Elephant
  Elephas maximus
Chital
  Axis axis
Sambar
  Cervus unicolor
Leopard
  Panthera pardus
Jackal
  Canis aureus
Hog Deer
  Axis porcinus
Sloth Bear
  Melursus ursinus
Rhesus Monkey
  Macaca mulatta
Hyena
  Hyena hyena
Gavial
  Gavialis gangeticus
Mugger
  Crocodylus palustris
Black-necked Stork
  Xenorhynchus asiaticus
Red Jungle-fowl
  Gallus gallus
Black Partridge
  Francolinus francolinus

Visiting: The Park HQ at Dhikala is 150 miles by road from Delhi and there is a railway to nearby Ramnagar. Tracks facilitates exploration and elephants can be hired. There are observation towers.

Hidden by the long grass of Corbett National Park lives one of the largest concentrations of chital, the beautiful spotted deer of India. Seen best from the back of an elephant, these animals are exceedingly difficult to photograph.

Coto Doñana, Spain

The Coto Doñana is a place for birds, with an impressive collection of waterbirds and raptors. It also provides a refuge for the largest remaining population of the pardel or Spanish lynx (above). Purple herons (right) nest nearby in the safety of the reed beds.

One of Europe's top three wildlife zones lies at the mouth of the River Guadalquivir where the mighty river reaches the open Atlantic Ocean. Here the sea has built up a huge system of dunes, over a hundred feet high in places, that diverts the river southwards and turns what would have been a huge estuary into a maze of marshes. The floods of winter are halted in their tracks by the Atlantic swell, and fresh water floods an area of several hundred square miles called the Marismas. Wild horses and cattle roam these splashy marshes along with a wealth of birds that it would be hard to equal in Europe.

The Coto Doñana Reserve was purchased with the aid of the World Wildlife Fund and is administered from offices in Seville. Though properly applied only to the coastal belt, the whole area of the Guadalquivir delta is frequently referred to as the Coto. In its search for new holiday resorts, the tourist industry is developing the coastal dunes at an alarming rate but the hinterland remains unspoilt and vitually untouched. Umbrella and stone pines cover the landward dunes and salicornia scrub the drier edges of the Marismas. The trees provide a home for the predators that find an easy living on the nearby marshes. In particular they are the last stronghold of the Spanish Imperial eagle and the pardel lynx. The eagle is easily seen, but the lynx is more cautious and most visitors must content themselves with a view of this exciting animal prowling its cage in the grounds of the reserve's headquarters, the Palacio de Doñana.

As the Marismas dry out, aquatic birds concentrate in fewer and fewer areas. Little and cattle egrets, night and squacco herons gather in isolated cork oaks to breed by the thousand. Grey and purple herons, though colonial, are more scattered and less obvious. The first islands to dry out (they are called vetas), are invaded by hordes of plovers, terns and gulls and by the late migrant pratincoles that scream overhead on sickle-shaped wings.

Because of its southerly position the Coto Doñana has interesting African elements among its fauna including white-headed duck, marbled teal and the secretive and timid purple gallinule—the latter is a distinct species from the American bird of the same name.

Overhead soar the scavengers and raptors. Three species of vulture, the black, griffon and Egyptian, gather at carcasses, while black kites haunt the

Birds are everywhere on the marismas and lagoons of the Coto Doñana Reserve, but the spectacle is dominated by the regular flights of cattle egrets (right) and night herons (left) that change over, shift fashion, every evening from roost to feeding ground.

Species of particular interest

Pardel Lynx
  Felis caracal
Red Deer
  Cervus elaphus
Fallow Deer
  Dama dama
Imperial Eagle
  Aquila heliaca
Short-toed Eagle
  Circaetus gallicus
Bonelli's Eagle
  Hieraetus fasciatus
Griffon Vulture
  Gyps fulvus
Egyptian Vulture
  Neophron percnopterus
Red Kite
  Milvus milvus
Black Kite
  Milvus migrans
Marsh Harrier
  Circus aeruginosus
Purple Gallinule
  Porphyrio porphyrio
Marbled Duck
  Anas angustirostris
Cattle Egret
  Ardeola ibis
Night Heron
  Nycticorax nycticorax
Purple Heron
  Ardea purpurea
Red-necked Nightjar
  Caprimulgus ruficollis

river banks and drying out floods in search of stranded fish. Red kites soar over the drier areas, while short-toed eagles search out snakes among the dune slacks.

Red and fallow deer scamper away through the scrub disturbing some of the hundred or so small bird species that include the gaudy hoopoe and the tricky little warblers—Dartford, Sardinian, spectacled and subalpine. Azure-winged magpies flit among the pines and it is strange to think that if they are not seen in Iberia the travelling naturalist would have to go to China to stand another chance of seeing them. Red-necked nightjars, secretive and still during the day, call and display at dusk when Scop's owls join in with their monotonous 'poop-poop' call.

Visiting: Easily reached from Seville via Almonte and El Rocio, where you should stop for herons and African specialities.

The track to the Palacio leads off this road to the left and is barred. Visiting permits must be obtained in advance from Estación Biologica de Doñana, Paraguay 1, Seville, Spain. Accommodation available for limited number of guests at the Palacio. Take food for the cook to prepare. Camping by permit. Car is essential and can be hired in Seville.

Cradle Mountain and Lake St Clair National Park, Australia

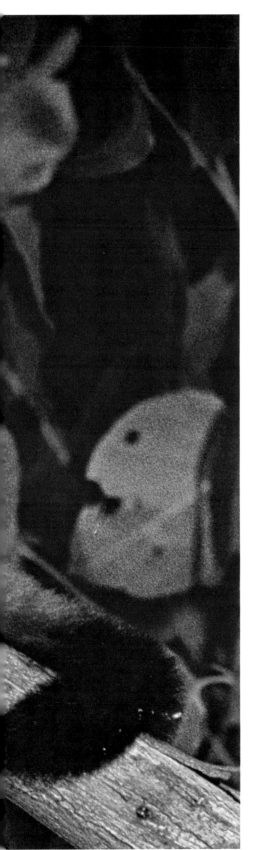

Nestling among the highlands of Tasmania lie the Cradle Mountains, home of the progressively more elusive echidna or spiny anteater (below). It is becoming very difficult to see this animal on the Australian mainland. Adapted to a life among the trees, the sugar glider (left) has extendable flaps of skin that enable it to sail-plane from tree to tree.

One of the three, and by far the largest of Tasmania's national parks, Cradle Mountain is situated in the highlands of that island. Here latitude and altitude combine to produce an almost 'European' climate that is cold but sunny in spring, summer and autumn and covered with snow during the winter. Cradle Mountain lies in the north-west of Tasmania with Mount Ossa the highest peak of the island. Lake St Clair is used as a reservoir and is surrounded by pine and eucalyptus forests together with plains and savannah. The higher slopes exhibit Australian alpine vegetation and there are many glacial lakes and other evidence of glaciation. This is typical, rugged mountain scenery and has suffered little or no interference by man.

The Park is known chiefly for its abundance of mammals which have become attracted to the overnight hut areas. Being an offshore island, and a temperate one at that, Tasmania does not have the wealth of species of the Australian mainland, but several of the twenty-odd Tasmanian endemic birds are found within the Park boundaries.

Red-bellied melon, red-necked wallaby, Tasmanian brush-tail possum, long-nosed rat kangaroo can all be seen with some degree of regularity here. Marsupial mice appear around the huts in the late

afternoon and the gliding possums are
usually in evidence at that time. The
eastern native cat, perhaps better called
the quoll, is now virtually confined to
Tasmania, and it is quite possible that the
Tasmanian tiger wolf or thylacine is still
hiding away somewhere in this area.
It has not been seen or properly
identified since about 1950 but there are
all sorts of rumours about footprints and
blood-chilling calls in the night. The area
for investigation is the west of the Park.
On surer ground the duck-billed platypus
can be found by those that explore
thoroughly. Armed with a fearsome
battery of spines, the spiny anteater or
echidna can also be seen here and there
throughout the area.

Birds include the introduced superb
lyrebird and such species as the boobook
owl, owlet nightjar, kookaburra, several
parrots and cockatoos, ground thrush, etc.

Visiting: By road from Hobart 105 miles
away. Use the Tarraleak Highway via
Derwent Bridge to Lake St Clair from the
south; and via Sheffield and Wilmot to
Waldheim in the north. Waldheim has
chalets and cabins, Lake St Clair camping
areas. There is a 5-day 52-mile trek
between the two areas—the only way
really to see the wildlife—with free huts
for overnighting. From St Clair a day can
be saved by crossing the lake by boat.

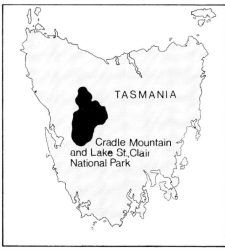

*Red-necked wallabies (above) inhabit the
more open areas of the Cradle Mountains,
whereas the thylacine, or Tasmanian tiger
(right), is one of the most elusive animals.
Seldom seen from one year to the next,
if it exists at all it is in this mountainous
area of central Tasmania.*

TASMANIA

Cradle Mountain
and Lake St. Clair
National Park

Species of particular interest

Tasmanian Devil
  Sarcophilus harrisi
Native Cat
  Satanellus hallucatus
Platypus
  Ornithorhynchus anatinus
Echidna
  Tachyglossus setosus
Wombat
  Vombatus ursinus
Red-bellied Melon
  Thylogale billardierii
Red-necked Wallaby
  Protemnodon rufogrisea
Tasmanian Brush-tail Possum
  Trichosurus fuliginosus
Long-nosed Rat Kangaroo
  Potorous tridactylus
Marsupial Mouse
  Antechinus minimus
Sugar Glider
  Petaurus breviceps
Tasmanian Tiger (extinct?)
  Thylacinus cynocephalus
Masked Owl
  Tyto novaehollandiae
Boobook Owl
  Ninox novaeseelandiae
Owlet Nightjar
  Aegotheles cristatus
Painted Quail
  Turnix varia
Superb Lyrebird
  Menura novaehollandiae
Yellow-tailed Black Cockatoo
  Calyptorhynchus funereus funereus
Green Rosella
  Platycercus caledonicus
Rufous Fantail
  Rhipidura rufidorsa
Golden Whistler
  Pachycephala pectoralis
Black Carrawong
  Strepera fuliginosa
Kookaburra
  Dacelo gigas
Yellow Wattle-bird
  Anthochaera paradoxa
Scrub-tit
  Acanthornis magnus
Brown Scrub-wren
  Sericornis humilis

Danube Delta, Romania

Europe's longest river runs across a continent, through half a dozen countries, before dumping a hundred million tons of silt a year at its mouth. Here, on the tideless Black Sea, the silt has created a huge delta, one of the gems of the European wildlife collection. Below Tulcea the Danube divides into three major arms—the Chilia, Sulina and the St Gheorghe. The Sulina is the main artery of communication and has been straightened and improved to facilitate passage of sea-going ships deep into the heart of the continent. The other channels remain more natural. Each is lined with thick hardwood woodland, established where the silt has formed banks along the paths of the river's flow. Backing the narrow bands of woodland stretch vast marshes of floating reeds and

The largest of European wetlands, the Danube Delta, provides a home for vast legions of water birds. Little egrets (left) inhabit almost every pool, while the abundant supply of fish is sufficient to make this the only breeding site in Europe for the great white pelican (below).

shallow open water. And everywhere there are birds.

Apart from sturgeon—most Romanian caviare is exported—the delta is alive with fish, huge fish in huge quantities. No doubt this is the reason why it is the only regular breeding site in Europe for the white pelican. These superb birds circle on great stiff wings or glide, flapping in turn, into some remote lagoon. They seem too exotic for Europe. Four thousand birds can be found in the Danube Delta, but there are also a few hundred Dalmatian pelicans, which breed in Greece and other parts of the Balkans as well as in the delta.

Glossy ibises also breed nowhere else in Europe, but they can always be seen— they're everywhere. Five thousand, ten thousand, who knows? Just vast numbers of birds in every direction.

The delta provides a home for numbers of otherwise rare great white egrets— easily seen and approached. Little egrets, night, squacco and purple herons search for fish along the banks of the smaller channels that form a maze among the marshes. Pygmy cormorants are rare elsewhere but common enough in the Danube Delta where they easily out- number the common cormorant.

Duck and waders are there in plenty with ferruginous as common as any. Ruff and black-tailed godwits can be seen by the hundred, and white-winged, black and whiskered terns hawk insects from the surface.

To the south lie the two subsidiary lakes—Sinoe and Razelm. They are huge expanses of water but as they dry out they provide a home for a host of birds including the scarce ruddy shelduck on Popina Island, which is a strictly protected reserve. Where the two lakes meet vast numbers of gulls, terns, pratincoles and greylag geese can be found. Herons and little bitterns, as well as off-duty pelicans, are usually to be seen.

Visiting: Take a regular steam-boat from Tulcea to Maliauc where there is a hotel. Small rowing boats, with expert boatmen, may be hired to explore the backwaters and inaccessible lagoons, but larger boats cater for parties organized in Tulcea.

Rare elsewhere in Europe great white egrets haunt almost every backwater of the marshes of the Danube Delta where they fish patiently with curiously kinked necks.

Species of particular interest

White Pelican
 Pelecanus onocrotalus
Dalmatian Pelican
 Pelecanus crispus
Glossy Ibis
 Plegadis falcinellus
Greylag Goose
 Anser anser
Great White Egret
 Egretta alba
Pygmy Cormorant
 Phalacrocorax pygmaeus
Little Egret
 Egretta garzetta
Night Heron
 Nycticorax nycticorax
Squacco Heron
 Ardeola ralloides
Ferruginous Duck
 Aythya nycora
Whiskered Tern
 Chlidonias hybrida
Caspian Tern
 Hydroprogne tschegrava
Pratincole
 Glareola pratincola

Etosha National Park, South-west Africa

Etosha lies 260 miles north of Windhoek, the capital of South-west Africa. When it rains, over 3200 square miles are flooded to a depth of about three feet. This pan is, in fact, the bed of a former lake fed by the Kunene River, but for some reason unknown the river altered its course leaving Lake Etosha to dry out under the fierce African sun. Now it is a wilderness of flat, low-lying land that every so often becomes an incredibly rich wetland.

Within its 16 million acres Etosha National Park covers part of the coastal Namib Desert—a desert created by the cold offshore waters of the Benguela Current that sweep northwards from the Antarctic seas. Gradually the landscape becomes progressively less arid as one moves eastwards, changing from bare dunes with little but gemsbok to recommend them, to the open thornbush and savannah with its great herds of antelopes and their attendant predators and scavengers.

Until late last century this wilderness was untouched, the home of a few nomadic Bushmen and Ovambos living in balance with the herds of game. An outbreak of rinderpest at the turn of the century destroyed these herds and the wandering tribesmen moved elsewhere in search of game—Etosha became a waste-land, the empty quarter of an empty quarter. Governor Lindquist immediately proclaimed it a game reserve—as good a way of administering an area as any. Only since 1952 has the government begun a programme of exploitation and commercialization. The fort at Namutoni, once the scene of a siege and dramatic escape, has been rebuilt and converted into one of the most attractive and luxurious safari hotels to be found anywhere.

The Park lies some 4000 feet above sea level and enjoys an average rainfall of some seventeen inches. The heat in summer can be phenomenal and quickly dries out the flash flooding and the natural waterholes. To overcome water shortage the Park authorities have installed artificial supplies and these act as a great draw and attraction to the animals. A road system has been designed to pass every drinking station and notable waterhole enabling the visitor to see the largest herds of antelope in southern Africa in close-up and comfort.

Springbok, zebra and wildebeest are the most numerous of the large animals and the attractive kudu are more common here (in the south-east) than in most other African parks. Giraffe and elephant

The dry, arid wastes of the Kalahari are ideally suited to the gemsbok (left), a near relative of the oryx. The more fertile areas of grassland provide a home for a wider variety of game, including the high-jumping springbok, (below), seen here in company with a wildebeeste or gnu.

are widespread though the black rhinoceros is rare and found only in the west. Damara dik-dik can be found at Namutoni near the lodge.

Cats have created a considerable problem at Etosha. When the herds of game migrate many lions in particular are left with little choice but to leave the Park and feed on the domestic animals that graze so invitingly outside. A wire fence stops many but inevitably some lions manage to get through. Naturally the local farmers defend their herds. Thus what will doubtless become a more commonplace problem is enacted at Etosha–the conflict between Park and agriculture.

Visiting: By road from Windhoek or by air to airstrips at the camps and centres in the Park. There are centres at Okakuejo, Halali and Namutoni, each offering varied standards of accommodation. Each has a swimming pool, garage and camp site.

Species of particular interest

Lion
  Panthera leo
Leopard
  Panthera pardus
Cheetah
  Acinonyx jubatus
Black Rhinoceros
  Diceros bicornis
Gemsbok
  Oryx gazella
Kudu
  Tragelaphus imberbis
Springbok
  Antidorcas marsupialis

Hartmann's Mountain Zebra
  Equus zebra hartmannae
Elephant
  Loxodonta africana
Impala
  Aepyceros melampus
Red Hartebeest
  Alcelaphus caama
Eland
  Taurotragus oryx
Steenbuck
  Raphicerus campestris
Damara Dik-dik
  Rhynchotragus kirki
Giraffe
  Giraffa camelopardalis

Hyena
  Hyena hyena
Flamingo
  Phoenicopterus ruber
Secretarybird
  Sagittarius serpentarius

In the dry season game and predators alike find water around the fertile surrounds of Etosha Pan. (Opposite) Greater kudu are common, but must be ever watchful for the powerful, bone-cracking jaws of the spotted hyena (below)—a much maligned species. Both are accompanied by birds, the hyenas by the scavenging marabou stork, the kudu by tickbirds.

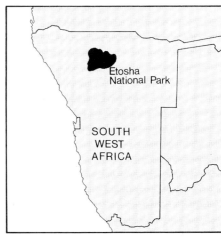

Etosha National Park

SOUTH WEST AFRICA

Everglades and the Keys, Florida, USA

The Everglades of southern Florida are justly famed as a refuge for wildlife. Few other areas can match the vastness and richness of this incredible complex of marshes, ponds and mangrove swamps. Extending for over a hundred miles, the southern part of the Everglades have been included in the National Park of that name. Elsewhere there is immense pressure on the land that can be converted to extending the happy holiday atmosphere of Miami Beach and all that.

Like the Everglades the Florida Keys gain a considerable part of their attraction from their uniqueness within the boundaries of the United States. They are lush and sub-tropical, gleaning a number of species (of birds in particular) that extend no further, or which are exceedingly rare, northwards. Connected one to another by old railroad bridges the Keys are a chain of islands extending south-westward from the mainland coast—they are part of the same system that created Miami Beach from the sea. United States Route National Number One connects the islands providing a quite unique opportunity for even the most casual visitor to see wildlife. In many cases it is not even necessary to leave the car.

Florida is superb for wildlife and while Loxahatchee, where fewer than twenty Everglade kites are making their last North American stand, and Merritt Island, the last spot for the dusky seaside sparrow, are outside our area, the major Florida specialities are down here in the south. Key deer, the smallest form of the white-tailed deer, find sanctuary in the National Key Deer Refuge established in 1954. There are now more than three hundred of these attractive animals. Likewise the great white heron has a refuge named after it, which covers some 2000 acres, most of which overlaps the Key Deer sanctuary.

Out on the Keys, as on all islands, the bird species become fewer while their numbers become greater. As the traveller progresses outwards he will note a variety of waterbirds—brown pelicans sit on buoys and posts, frigatebirds chase overhead though they do not breed, roseate spoonbills feed in the creeks along with reddish egrets, Louisian herons and double-crested cormorants. Swallow-tailed kites glide by, surely one of the world's most graceful birds, while the white-crowned pigeon delights the ornithologist by being found nowhere else in the USA.

While Key West, with its tropical

Bald cypresses (above) surround an Everglade backwater alive with wildfowl, while along the deeper channels that curious 'mermaid' the manatee (right) fights a losing battle with power boats and outboard engines. The area is one of the last strongholds of these aquatic mammals in North America.

attractions, is the end of the road, most wildlife enthusiasts will want to continue to Dry Tortugas—seven small, sandy islands lying some seventy miles to the west. Fort Jefferson stands as a monument on Garden Key while the adjacent Bush Key is alive with a hundred thousand or so sooty terns and a few hundred noddy and roseates.

Among the Everglades on the mainland it is the marshland birds that catch the attention. On White Water Bay, for example, almost all of the herons and egrets of the southern United States gather to breed in gigantic heronries. Great blue heron, snowy egret, little blue heron, green heron, black-crowned night heron, least bittern can all be found along with the otherwise scarce wood ibis. The limpkin, that strange heron-ibis-wader is common, and purple gallinules are somewhat less secretive than their European cousins. Bald eagles are probably as numerous as anywhere else in the country, though that doesn't say very much for them.

The Everglades and Florida Keys are not only a home of birds; there are alligators and crocodiles along the waterways and green turtles haul out at some of the more remote keys and can often be seen.

At Key West the marine aquarium will also prove an attraction. Though fast disappearing there are several places where manatees, those strange mermaids, can still be seen, though the propellers of outboard engines are chugging out the death knell of this aquatic mammal.

Visiting: The Keys can be explored by road, but Dry Tortugas must be reached by charter boat from Key West. Boat exploration is laid on for visitors and accommodation is plentiful. Arrange accommodation well in advance.

Among the most curious of Everglade birds is the limpkin (above left), a bird so strange that it is placed in a family of its own. Its long, decurved bill is used to extract snails from their shells. The American alligator (above right) is less specialized in diet and thrives under the protection afforded it from manufacturers of handbags and shoes.

Species of particular interest

Manatee
  Trichechus manatus
Raccoon
  Procyon lotor
Opossum
  Didelphis marsupialis
Alligator
  Alligator mississipiensis
Green Turtle
  Chelonia mydas
Swallow-tailed Kite
  Elanoides forficatus
Everglade Kite
  Rostrhamus sociabilis
Bald Eagle
  Haliaetus leucocephalus
Anhinga
  Anhinga anhinga
Brown Pelican
  Pelecanus occidentalis
Frigatebird
  Fregata magnificens
Great White Heron
  Ardea occidentalis
Great Blue Heron
  Ardea herodias
American Egret
  Casmerodius albus
Snowy Egret
  Egretta thula
Louisiana Heron
  Hydranassa tricolor
Roseate Spoonbill
  Ajaia ajaja
Little Blue Heron
  Florida coerulea
Green Heron
  Butorides virescens
Black-crowned Night Heron
  Nycticorax nycticorax
Least Bittern
  Ixobrychus exilis
Wood Ibis
  Mycteria americana
Limpkin
  Aramus guarauna
King Rail
  Rallus elegans
Purple Gallinule
  Porphyrula martinica
Sooty Tern
  Sterna fuscata
Noddy Tern
  Anous stolidus
White-crowned Pigeon
  Columba leucocephala
Mangrove Cuckoo
  Coccyzus minor

Galapagos Islands, Ecuador

The Galapagos lie six hundred miles off the South American coast and in that lies their uniqueness and importance. When Charles Darwin visited the islands in 1835 aboard HMS *Beagle* the theory of evolution that he had been wrestling with was one step nearer its formulation. But the Galapagos are no mere shrine to Darwin or his earth-shattering theory; they are a living laboratory and a magnificent place to see wildlife.

Situated on the Equator but cooled by the final fling of the cold Humboldt Current, the Galapagos are the remnants of submerged volcanoes. Yet the islands are only about a million years old. They rose from the sea lifeless, like a tropical Surtsey. It is the combination of their remoteness and their newness that make the Galapagos unique. When a new island

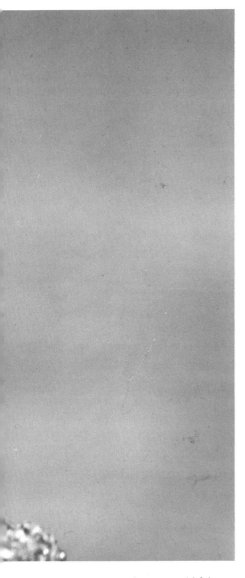

rises from the sea it is barren and lifeless. Gradually plants, animals and birds colonize and evolve in different ways to adapt to the circumstances in which they find themselves. The first colonists will be the seabirds that will simply find it a good spot to breed, but they will be followed by other marine creatures like turtles and, in the case of the Galapagos, marine iguanas.

Now because of their isolation the animals that have established themselves on the new island will tend to occupy varying niches, species will begin to diversify and perhaps even form new species. This has happened on the Galapagos with the endemic finches—Darwin's finches. Though sharing a common ancestor the finches have evolved into thirteen different species.

Some are like warblers and have evolved a thin picking bill, others have become nutcracking specialists, while one has evolved the habits, but not the tongue, of a woodpecker.

Also because of their isolation the species that have managed to colonize the new island will be insulated from competition and thus be able to survive—possibly even better than in the area whence they originally came. This must be true of the marine iguana for example.

The Galapagos then share the features of other islands that are 'new land' and isolated, but they also have a certain uniqueness that comes, in part at least, of their geographical position. The richness of the surrounding waters has brought a wealth of life to the islands. Penguins and albatrosses, for example, breed nowhere else on the Equator. Both are numerous and very confiding. Turtles, particularly the green turtle, are still common. The sealion is clearly a descendant of the California sealion, while the fur seal was taken to the verge of extinction by the commercial sealers of last century.

The giant tortoise, formerly common on ten or eleven islands, was hunted for food by passing ships and is now found only on two islands where it is strictly protected. In fact it is so well protected that it is difficult to find one without a number painted on its shell.

But above all the Galapagos birds will thrust their attention upon the visitor. All are extraordinarily tame, from the three species of boobies, whose name derives from their apparent stupidity in the face of danger, to the frigatebirds which sit on their nests and allow their portraits to be taken. Many birds are found nowhere else in the world—the albatross and penguin come to mind along with the finches—but there is a flightless cormorant, a hawk that would be called a buzzard in Europe and a mockingbird that is particularly fond of other birds' eggs and does not seem to mind being watched in the act.

Visiting: Regular air and ship services and many operators run natural history tours. There are tours by a luxury yacht based in the islands. Accommodation can be obtained on some islands, and the Charles Darwin Research Station at Academy Bay, Santa Cruz, copes with a few serious scientists.

A young Galapagos sealion (right) sits forlorn, weepy-eyed along the shoreline. Inland a pair of waved albatrosses (above) engage in a ritualized greeting ceremony. Note that both birds have been ringed as part of a scientific investigation.

Species of particular interest

Galapagos Sealion
  Otaria byronia
Galapagos Fur Seal
  Arctocephalus australis
Giant Tortoise
  Testudo elephantopus
Green Turtle
  Chelonia mydas
Land Iguana
  Conolophus subcristatus
Marine Iguana
  Amblyrhynchus cristatus
Lava Lizard
  Tropidurus albemarlensis
Galapagos Penguin
  Spheniscus mendiculus
Waved Albatross
  Diomedea irrorata
Flightless Cormorant
  Nannopterum harrisi
Blue-footed Booby
  Sula nebouxii
Masked Booby
  Sula dactylatra
Red-footed Booby
  Sula sula
Galapagos Hawk
  Buteo galapagoensis
Red-billed Tropicbird
  Phaethon aethereus
Lava Gull
  Larus fuliginosus
Vermilion Flycatcher
  Pyrocephalus rubinus
Yellow Warbler
  Dendroica petechia
Mockingbird
  Neosomimus spp.
Darwin's Finches
  Geospiza spp.

Glacier National Park, Montana, USA, with Waterton Lakes National Park, Canada

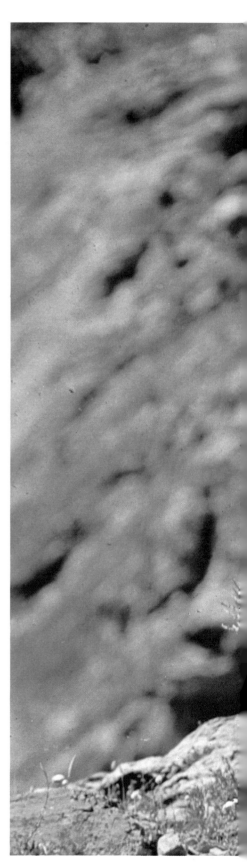

Named after the huge Illecillewalt Glacier, ten square miles of ice which actually lie in Canada, Glacier National Park covers 1600 square miles of upland country on the Canadian-United States border. It varies in altitude from 3200 to over 11,000 feet and though primarily a great scenic area with mountains and ice forming a backdrop to lakes set among the conifer woodlands, there is much wildlife to be seen. Almost all the large mammals that were formerly found so widespread in the American west can still be found here—only the bison is missing. Even this animal can be seen to the south in the National Bison Range which was established in 1908 and where the hills and canyons support a herd of up to five hundred animals.

As its name implies glaciers have made this landscape. Huge U-shaped valleys, often with lakes, carve deep into the mountains with subsidiary hanging valleys high up the hillsides. These valleys are covered with forests of lodgepole pine, Douglas fir, larches, pines and juniper. Among them can be found moose and beaver, both so typical of the American north-west. Black and grizzly bears hide away in the hills, and both are potentially dangerous. Elk and white-tailed deer are present in substantial herds and mule deer too are frequently seen. On the higher land the pronghorn antelope and bighorn sheep find the living ideal, while this is also the beat of the golden eagle soaring on spread wings over the rises and round the spurs.

A glacier-carved lake (left) occupies a valley deep in the interior of Glacier National Park, Montana. Below, bighorn sheep roam the rocky crags high up among the mountain tops. The clash of their horns during the rutting season may be heard up to a mile away.

The birds of the area are nothing special but typical of the eastern Rockies. Red-tailed and rough-legged hawks, prairie falcons and a variety of owls form the major predator force while the woods echo to the calls of nuthatches, western tanagers, Clark's nutcrackers and blue grouse. There are many warblers and chickadees and a chance of four distinct species of hummingbird. There are few shorebirds, though Wilson's phalarope breeds along with spotted sandpiper.

Visiting: There are hotels, motels and camp sites within the Park operated to enable visitors to get the best from this wilderness area. Particularly good for riding over or back-packing. Plenty of guided tours, and a road runs through the centre of the Park between St Mary and West Glacier. There are boat trips on some of the larger lakes.

Mule deer (above) are among the most common of the large animals found through the Rocky Mountain system. Much sought after by hunters, they are safe within the boundaries of the Park. Blue grouse (right) haunt the woodlands of Glacier; their elaborate courtship on 'lekking' grounds can be seen by any early riser.

Species of particular interest

Black Bear
  Ursus americanus
Grizzly Bear
  Ursus arctos
Coyote
  Canis latrans
Wolf
  Canis lupus
Bighorn
  Ovis canadensis
Porcupine
  Erethizon dorsatum
Snowshoe Hare
  Lepus americanus
Beaver
  Castor fiber
Pronghorn
  Antilocapra americana
White-tailed Deer
  Odocoileus virginianus
Mule Deer
  Odocoileus hemionus
Elk
  Cervus canadensis
Moose
  Alces alces
Wolverine
  Gulo gulo
Western Grebe
  Aechmophorus occidentalis
Harlequin Duck
  Histrionicus histrionicus
Red-tailed Hawk
  Buteo jamaicensis
Rough-legged Hawk
  Buteo lagopus
Golden Eagle
  Aquila chrysaetos
White-tailed Ptarmigan
  Lagopus leucurus
Blue Grouse
  Dendragapus obscurus
Ruffed Grouse
  Bonasa umbellus
Great Horned Owl
  Bubo virginianus

Gran Paradiso and Vanoise National Parks, Italy and France

High up in the Alps lie two of Europe's finest national parks, the Gran Paradiso in Italy and the Vanoise in France. The two share a common boundary along the Franco-Italian border and have a common fauna and flora. Though a few roads penetrate the parks they are lonely mountainous places, the preserve of a few cattle that graze them in summer, and the hill walker and naturalist. The boundaries are carefully drawn to exclude habitation and only a few mountain refuges provide accommodation. Several large glaciers help to maintain this rugged image, though ski resort developers have in recent years cast envious eyes at the high mountain slopes.

The fauna is alpine in character but at the park boundaries there are meadows and large coniferous woodlands clinging to the hillsides. Here small birds like redstarts, crested tits and red-backed shrikes find a niche, while along the streams grey wagtails and dippers are found. But the main *saison d'être* of the parks and their prime claim to fame is the home that they provide for ibex. These wild goats are now decidedly rare in Europe and for them the Vanoise and Gran Paradiso provide a safe refuge. They perch precariously on the rocky pinnacles before scurrying away across what seem to be sheer rock faces or vertical ice fields with a sure-footedness typical of their clan. There are several hundred ibex here, though seeing them does require at the very least a strenuous mountain walk.

Chamois, the little deer with turned over antlers, are also high altitude animals, though they come down among the trees in winter and on chilly summer nights. During the day they frequent the mountain sides, never penetrating as high as the ibex. They are equally as sure-footed and probably more numerous.

No one can visit the Alps without being aware of marmots. Their shrill whistles echo around the hillsides, but they are often difficult to see. Like most mammals they can see you before you see them and scurry underground into their burrows to safety. Marmots form one of the principal items in the diet of the golden eagle, found in both parks in good numbers. Elsewhere in these countries eagles have suffered serious persecution but they find a sanctuary in these national parks. Other birds include hazel and willow grouse and ptarmigan among the grouse tribe. Wallcreepers can be found on the sheer cliffs but, as elsewhere, they are difficult to see. Crested tits frequent

the forests and citril finches are found where the trees give way to open, more hostile areas. Snow finches and alpine accentors find a home on the highest mountains, but the visitor may be able to see them without too much effort from the highest passes. Like other Alpine animals they come lower in winter. Both yellow-billed and red-billed choughs soar over the crags in large parties.

Visiting: Access is without formality, but very few roads cross the parks proper.

Species of particular interest

Ibex
  Capra ibex
Chamois
  Rupicapra rupicapra
Marmot
  Marmota spp.
Golden Eagle
  Aquila chrysaetos
Hazel Hen
  Tetrastes bonasia
Buzzard
  Buteo buteo
Crested Tit
  Parus cristatus
Red-backed Shrike
  Lanius collurio
Snow Finch
  Montifringilla nivalis
Alpine Accentor
  Prunella collaris
Citril Finch
  Serinus citrinella

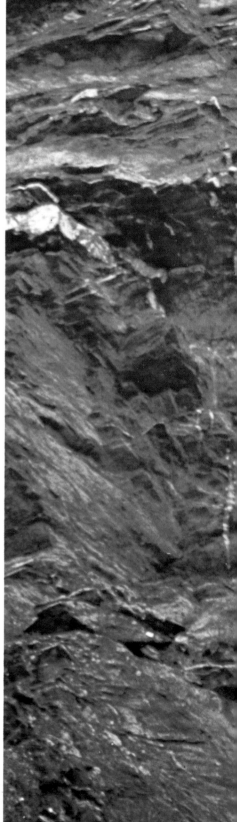

Widespread through the mountainous areas of the Old World, ibex are numerous among the rocky crags of Italy's Gran Paradiso National Park. No respecters of national boundaries, they are also protected in the adjoining French park, the Vanoise.

Henri Pittier National Park, Venezuela

Described as 'perhaps the most important refuge in Latin America' by one of the world's top bird men, Henri Pittier National Park covers 220,000 acres of the coastal range of mountains west of Caracas. Varying in height from sea level to 7690 feet the Park is a unique collection of habitats.

Inland, to the south, the jungle-clad hills give way to the interior basin of Venezuela which eventually becomes the Llanos. To the north is the arid area of the coast with cacti, while between the two is the Cordillera de la Costa (isolated from the main body of the Andes) with its more open areas of woodland. All these habitats can be explored via a road that transects the Park before ending at the beach. In the mountains it is altitude that determines the type of vegetation and thus the fauna.

In the heart of the Park lies the nature reserve of Rancho Grande, which is strictly preserved for nature and research. It is a high level area of cloud forest and boasts many of the most fascinating and obscure birds of the Park itself. In particular it is good for the woodcreepers, foliage gleaners, antbirds and flycatchers that are such a feature of the continent. Perhaps nowhere else can they be seen to such advantage. The bearded bellbird, the monotypic swallow-tanager, and a wealth of other tanagers makes for splendid, if confusing, bird-watching. Above all hummingbirds flit here and there—perhaps the most exciting group of birds in the world for those that simply want to watch and enjoy them.

Some 530 species of birds are found in Henri Pittier National Park—that's about the same number as has ever visited Britain and Ireland. Most are small passerines but there are also white hawks, ornate hawk-eagles, chachalacas, guans and a few pigeons.

Mammals are rather thin on the ground, though howler monkeys are difficult to miss with their shrill screaming. Three-toed sloths can sometimes be seen clambering about the trees. But above all this is a place for birds and one of the easiest of access in South America. Over 40,000 people visit it every year and there are no restrictions on access.

Just along the coast lies the excellent lagoon of Chichiriviche, one of the best waterbird areas in the Caribbean. It features thousands of flamingoes with hundreds of scarlet ibis and roseate spoonbills scattered among them. A wealth of other large water birds will make the foreign visitor feel more at home after the exotic experience of the birds of the National Park. Gulls, terns, boobies, pelicans, ibis, duck—all can be seen along with the Everglade or snail kite and a number of wintering shorebirds from North America.

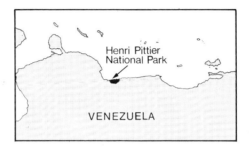

Visiting: Most visitors stay at one of the hotels in nearby Maracay and motor into the Park each day along the main road. The headquarters at Rancho Grande may be visited and guards may be able to act as guides.

Species of particular interest

Howler Monkey
  Alouatta spp.
Three-toed Sloth
  Bradypus tridactylus
White Hawk-eagle
  Spizastur melanoleucus
Rufous-vented Chachalaca
  Ortalis ruficauda
White-vented Plumelateer
  Chalybura buffonii
Booted Racquet-tail
  Ocreatus underwoodii
Collared Trogon
  Trogon collaris
Groove-billed Toucanet
  Aulacorhynchus sulcatus
Moustached Puffbird
  Malacoptila mystacalis
Bearded Bellbird
  Procnias averano
Swallow-tanager
  Tersina viridis
Glittering-throated Emerald
  Amazilia fimbriata
Magnificent Frigatebird
  Fregata magnificens
American Flamingo
  Pheonicopterus ruber
Roseate Spoonbill
  Ajaia ajaja
Scarlet Ibis
  Eudocimus ruber
Forster's Tern
  Sterna forsteri
Chestnut-fronted Macaw
  Ara severa

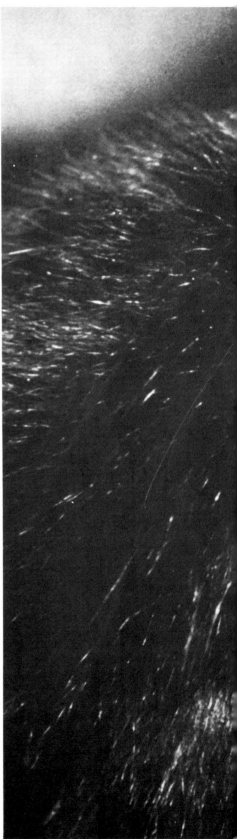

Red howler monkeys scream from the tree tops of the Henri Pittier Park, a haven for all kinds of South American wildlife in an otherwise hostile environment.

Heron Island Marine National Park, Australia

No wildlife paradise could conceivably be only a half a mile long by three hundred yards wide but so it is with Heron Island. This coral island is, in fact, just the tip of the iceberg of one of the best parts of the Great Barrier Reef, a reef that stretches for 1200 miles along the eastern coast of Australia. Some of the hundreds of islands that litter this great growth of coral are larger and boast a richer fauna, but none has the reputation for fish enjoyed by Heron Island.

In comparison with the underwater fauna, mammals and birds are decidedly few in species if not in numbers. With the centre of the island virtually a pisonia jungle life is confined to the beach and its environs. Here between October and April, leatherback and green turtles haul up to lay their eggs in one of the few places they are really safe. Also in October short-tailed shearwaters (muttonbirds) come ashore by the hundreds of thousands to nest in the burrows that virtually undermine the surface of the island. In the thickly covered centre of Heron Island white-breasted sea eagles find a safe nesting place ideally suited to their easy mode of living on the fish of the reef.

But visitors to Heron Island do not come with binoculars. They are equipped with mask and snorkel, and perhaps air-line, aqualung and wetsuit. They wish to see one of the greatest wildlife paradises on earth, the coral and fish of the Barrier Reef. It is difficult to pick out species from among the many hundreds that swarm over the coral outcrops (called bommies locally) around Heron Island. Butterfly fish are attractive and

Below, millions of fishes swarm among the staghorn coral of the Barrier Reef, originally the greatest underwater paradise on earth. Green turtles (left) glide effortlessly past, but also crawl out on the coral atolls to lay their eggs deep in the sandy beaches. Prized as a delicacy, turtles are in need of urgent protection.

intriguing—there are thirty species here each with its own distinctive markings and ecological niche. There are striped ones and plain ones, those with and without spots and of varied length of snout. Then there are the coral trout, the groupers, the harlequin tusk fish, the snappers, the sweetlips and the parrotfish. The latter have a hard 'bill' of teeth that is used to hack off coral to obtain the polyps inside. There are barracuda and trumpetfish, which Sir Peter Scott described as 'yellow submarines'—strange cigar-shaped fish with a deadly siphon device that sucks in small fish.

The bommies themselves are the haunt of cleaner wrasse. These have a strange relationship with a variety of other species that let the smaller fish clean them of parasites. The cleaners even enter the larger fish's mouth and sometimes emerge from its gills. But there are false cleaners too—gobies and blennies that look just like cleaners but take bites of their host as well as its parasites. Even the gigantic manta ray comes into the bommie for a clean up.

The various corals create a variety of formations from brain coral to the various sea fans. In some areas of the Barrier Reef the sixteen armed crown-of-thorns starfish has eaten its way through the coral leaving a trail of death and destruction in its wake. But the immense wealth of coral at Heron Island is a reassurance to many of the Reef's power to survive the onslaught of the starfish.

Fish, coral, molluscs, crustaceans, nudibranchs, turtles and birds in immense variety are offered by Heron Island, one of the gems of Australia's Great Barrier Reef.

Visiting: By helicopter or launch from Gladstone. There is simple accommodation on the island which must be booked in advance. Fresh water for washing is scarce.

Species of particular interest

Green Turtle
  Chelonia mydas
Leatherback Turtle
  Dermochelys coriacea
White-breasted Sea Eagle
  Haliaetus leucogaster
Short-tailed Shearwater
  Puffinus tennirostris
Oystercatcher
  Haematopus ostralegus
Noddy Tern
  Anous stolidus
Butterfly Fish
  Chaetodon spp.
Barracuda
  Sphyraena barracuda
Sweetlips
  Lethrinus chrysos
Coral Trout
  Serranidae
Manta Ray
  Manta birostris
Cleaner Wrasse
  Labroides dimidatus
Harlequin Tusk Fish
  Bodiamidae
Crown of Thorns Starfish
  Acanthaster planci

Australia's Barrier Reef is not only an underwater haven. Its islands and atolls support huge numbers of seabirds, including white-capped noddies that nest among the scattered trees.

Kabarega National Park, Uganda

Kabarega (formerly Murchison) National Park, covering 1557 square miles, is the largest in Uganda and one of the largest national parks in the world. It lies in the north-west of the country on either side of the great Victoria Nile and includes the confluence of that great river with its sister, the Albert Nile. In the centre of the park the Victoria Nile plunges through the narrow gap and over the cliff that gives the park its name.

While much of Kabarega consists of open plains of grass, it is the river that dominates the park and its fauna. There are areas of swamp, forest and woodland but no habitat makes the park what it is more than the Nile. For every visitor, be he scientist or tourist, the boat trip from the park headquarters at Paraa to the Kabarega (Murchison) Falls is sheer bliss. Crocodiles line the banks and take to the water as you pass, just as in the movies. This stretch of the Nile is now the most important area in the world for crocodiles and crucial to their survival in the wild. Demands for their skins for ladies' handbags and shoes have decimated the once common reptiles, and though they can be farmed there is still great pressure

on the wild population. Hippopotamus are also common here as in few other areas of Africa. The visitor travelling by river will see buffalo in fairly large herds and exceptional numbers of elephant. The plains of Kabarega are the home for truly vast numbers of elephant, perhaps rivalled in the world only by the elephants of Tsavo. Estimates vary between four and ten thousand animals. The other large animals, the white and black rhinoceros can both be found—the former, of course, is extremely rare and to be found otherwise only in a few areas like Garamba National Park in the Congo and Umfolozi Game Reserve in Zululand. It was introduced into Kabarega from the lands west of the Nile.

Kabarega is also home to a large variety of antelopes and gazelles like kob, oribi, three duikers, defassa waterbuck, reedbuck, bushbuck and Jackson's hartebeest. Lions and leopards provide the main predatory force, though several eagles account for the smaller prey.

The Rabongo Forest in the south-east adds considerably to the already large list of birds found in the Park, but is primarily known as the home of chimpanzees. Among the more unusual birds found here are the occasional shoe-billed and woolly-necked storks and sooty falcon, the strange bat hawk, palm-nut vulture, both pale and dark chanting goshawks, the delicious pallid harrier, Peter's finfoot and hundreds of others.

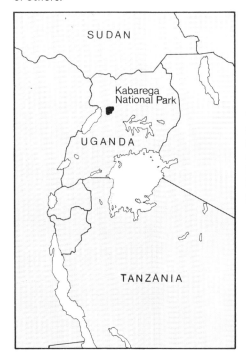

Visiting: The Park can be reached by road and air and there is an excellent network of roads to all parts of the area. Accommodation of the usual luxury can be found at Paraa Safari Lodge and Chobe Lodge. Camping is permitted in various parts of the Park.

While secretarybirds (left), so called for the quills 'behind their ears', stalk the dry grasslands, the river is the centre of attraction in Kabarega. Along its sandy banks, crocodiles (below) sun themselves. Much persecuted elsewhere, these huge reptiles find protection within the boundaries of the Park.

Species of particular interest

Hippopotamus
  Hippopotamus amphibius
Elephant
  Loxodonta africana
Buffalo
  Syncerus caffer
Black Rhinoceros
  Diceros bicornis
White Rhinoceros
  Ceratotherium simum
Rothschild's Giraffe
  Giraffa camelopardalis rothschildi
Jackson's Hartebeest
  Alcelaphus buselaphus jacksoni
Kob
  Kobus kob

Oribi
  Ourebia ourebi
Defassa Waterbuck
  Kobus defassa
Bohor Reedbuck
  Redunca redunca
Bushbuck
  Tragelaphus scriptus
Crocodile
  Crocodylus niloticus
Shoe-billed Stork
  Balaeniceps rex
Woolly-necked Stork
  Dissoura episcopus
Saddle-billed Stork
  Ephippiorhynchus senegalensis
Sacred Ibis
  Threskiornis aethiopicus

Secretarybird
  Sagittarius serpentarius
Palm-nut Vulture
  Gypophierax angolensis
Sooty Falcon
  Falco concolor
Chanting Goshawk
  Melierax poliopterus
Finfoot
  Podica senegalensis
Wattle Plover
  Afribyx senegallus
Abyssinian Roller
  Coracias abyssinica
Broad-billed Roller
  Eurystomus glaucurus
Ground Hornbill
  Bucorvus leadbeateri

Kanha National Park, India

Of all the wildlife areas of India none comes nearer the visual impact of an African safari than Kanha National Park. Though the centre of India is dominated by huge sal forests with dense under-growth and clinging creepers, here and there are areas of open parkland dotted with trees just like the acacia savannas of East Africa. Kanha is one of these clearings, and one of the largest. It is almost the only place in India where herds of large animals can be seen in the open, grazing the lush post-monsoon grass. Because of this openness it is also the only place in India where a sight of a tiger in daylight can be guaranteed—though no one actually 'guarantees' tigers anywhere.

Among the gems of Kanha is its large herd of barasingha (which means twelve-pointer), better known as swamp deer. These fine antlered deer are confined to the Indian sub-continent and found elsewhere only in the terai of southern Nepal. Unfortunately even here the barasingha are declining fast and may soon disappear. Chital, the beautiful spotted deer, is also declining rapidly, though blackbuck seem to have maintained their numbers at about thirty to forty animals.

The gaur, misleadingly called the Indian bison, is present in reasonable sized herds as are sambar—these large animals providing further opportunity for predators and tiger in particular. The Bengal tiger now numbers 1800 in the world. Being such a solitary animal, and a nomad as well, it may well become extinct in the wild by the end of the century. The Indian government is making a determined effort to save the tiger, but as always money is the problem. Kanha is the best place in India for tiger and yet it is a mere ninety-seven square miles in extent and enjoyed a budget of only $7500 in 1964-5. The tiger will not be saved in this way. Only by establishing truly large tiger reserves, where the animals can wander wide and far will this splendid creature be saved.

Kanha has not been well studied ornithologically but it has a list of about a hundred species. Among the more interesting are red jungle fowl (wild chicken), drongos, rollers, peacocks, parakeets etc.

Kanha National Park

INDIA

Visiting: This is not an easy place to get to. It is 60 miles from Jabalpur and 35 miles from Khana. There are two rest houses and elephants may be hired. The Park is open from November to June, though the best season is in April and May when the tracks are dry and the vegetation has died down.

Species of particular interest

Tiger
  Panthera tigris
Leopard
  Panthera pardus
Gaur
  Bos gaurus
Chital
  Axis axis
Sambar
  Cervus unicolor
Swamp Deer
  Cervus duvauceli
Blackbuck
  Antilope cervicapra
Sloth Bear
  Melursus ursinus
Hyena
  Hyena hyena
Wild Pig
  Sus scrofa
Adjutant Stork
  Leptoptilos dubius
Red Jungle Fowl
  Gallus gallus
Blue Peafowl
  Pavo cristatus
Black Drongo
  Dicrurus adsimilis
Indian Roller
  Coracias benghalensis

Among the swamps and pools of Kanha National Park one of the few herds in the world of barasingha, or swamp deer, can be found. They are easily approached on elephant back.

Kaziranga Wildlife Sanctuary, India

Kaziranga—home of the great Indian one-horned rhinoceros and, together with the reserve at Chitawan in Nepal, the area where the future of this great beast will be decided. Only a thousand are now left in the world, over half of which are split between Chitawan and Kaziranga. This Indian stronghold stretches for 25 miles along the southern bank of the Brahmaputra and covers a total of 166 square miles. Back in the 1930s the area was a Forest Reserve but also completely out of bounds, a vast malarial swamp where the concept of active management was unknown. Today it is one of India's wildlife show-places, a swamp land and forest where the trees and elephant grass provide ideal conditions for the rhinos.

Against a backdrop of the Himalayas the rhinos of Kaziranga are viewed from elephant back and are usually co-operative. The comparative openness of the reserve with large splashy areas covered with grass, plus the relatively high population of rhinos makes Kaziranga perhaps the best site for these animals outside Africa. In some of the more remote parts of the Sanctuary the rhinos are not used to seeing tourists and will often charge, but those that see elephant-borne people every day are generally docile and

There is no place on earth where the sight of a tiger can be guaranteed. But the visitor is in with a chance at the rich Kaziranga National Park in Assam. Though not a 'tiger reserve' the Park supports a good population of these magnificent cats (right). Wild Indian elephants are also becoming rare, but this old tusker (above) was found at Kaziranga.

approachable. Indian rhinos like to wallow every day and several small pools are known to hold them.

Kaziranga is also the home of wild elephant, another species that is fast disappearing from its native haunts. Frequently several hundred move down from the neighbouring Mikir Hills to the plains and wallows of the Brahamaputra.

Wild buffalo are another element in the fauna of large game animals that are found at Kaziranga. The herds are now small but in them lies the last chance of survival for another species. Wild pig and the sloth bear are also found, though as elsewhere, views of the latter are decidedly unusual. That fine but rare deer the barasingha or swamp deer is present as are hog deer and sambar.

Birds are numerous and varied and include numbers of the more familiar and spectacular large water birds. Black-necked and open-billed storks, pelicans, darters, both bronze-winged and pheasant-tailed jacanas, Indian white-backed and griffon vultures, all are found quite commonly.

As at Chitawan the greatest danger faced by the rhinos and other inhabitants of the Sanctuary is domestic cattle. It is extremely difficult to prevent inroads of stock every day and little that the authorities can do to prevent the extension of grazing and the decline of the rhinos. Poaching too is a problem, no fewer than thirteen rhinos were killed in 1966; but this is a danger that can and should be eliminated.

Species of particular interest

Indian Rhinoceros
  Rhinoceros unicornis
Wild Elephant
  Elephas maximus
Wild Buffalo
  Bubalis bubalis
Swamp Deer
  Cervus duvauceli
Sambar
  Cervus unicolor
Hog Deer
  Axis porcinus
Wild Pig
  Sus scrofa
Sloth Bear
  Melursus ursinus
Black-necked Stork
  Xenorhynchus asiaticus
Open-billed Stork
  Anastomus oscitans
Great White Egret
  Egretta alba
Intermediate Egret
  Egretta intermedia
River Tern
  Sterna aurantia
Black-bellied Tern
  Sterna acuticauda
Pheasant-tailed Jacana
  Hydrophasianus chirurgus

Visiting: Most visitors fly to Gauhati and then motor to the tourist lodge, where elephant trips explore the Sanctuary.

Much like domesticated stock, wild buffaloes can be found feeding among the marshes of Kaziranga in small herds. Their spread of horns (left) is larger and in general they look more healthy than the domestic beasts.

Keoladeo Ghana Wildlife Sanctuary, India

The trouble with giving areas new names is that they seldom stick. Thus the Ghana Sanctuary is generally referred to as simply as Bharatpur and then instantly recognized as the best bird sanctuary in India and one of the greatest wildlife areas in the world. Bharatpur lies west of Agra and no more than an hour's drive from that other wonder of the world, the Taj Mahal. Its eleven square miles include a large shallow lake that was once the private shoot of the Maharaja of Bharatpur. He still retains the winter shooting rights when fifteen species of duck descend by the thousand to feed among the rich overgrown waters. The record shoot was 4273 birds in a day in November 1938 when the slaughter was graced by Lord Linlithgow. Today such massacres are fortunately not indulged, though the number of duck seems not to have declined. Bar-headed and greylag geese are also present as is the interesting Indian pygmy goose. But above everything else the winter sees the rare Siberian white crane at Bharatpur. These exquisite birds nest only in one remote area of Siberia and migrate regularly only to this particular spot in India. Others may winter further west in Iran. These birds are really the Asiatic equivalent of the whooping crane of North America, and although not as rare as that species they are declining fast. Like all cranes they dance even in winter quarters and against the backdrop of

Keoladeo Ghana Wildlife Sanctuary

INDIA

Bharatpur that indeed is a fine sight.

Unfortunately the best time at Bharatpur is the summer when thousands of waterbirds gather to breed. The large, long-legged birds are the most obvious. Painted storks look ungainly as they cling to the top twigs of some tree but their plumage makes them one of the more attractive members of their family. Spoonbills and grey herons, little and intermediate egrets, purple herons and the strange-looking black ibis can all be found. Open-billed storks too parade the marshes of Bharatpur, their bills strangely adapted to their food. Indian darters swim among the vegetation and haul out like cormorants to dry their wings. Snake-bird is a very appropriate

The greatest waterbird site in Asia, the Keoladeo Ghana Sanctuary at Bharatpur, swarms with birds. Only the daily influx of cattle and domestic buffalo (left) spoil the scene. This is the stronghold of the great Siberian white cranes (below), which migrate from the north to winter among its shallow and rich lagoons.

alternative name for these serpentine necked birds.

Sarus cranes are numerous during the winter and several pairs stay on to breed. Because of their elaborate courtship they have become a symbol of a happy union in India and are protected by villagers. As a result they have become quite tame and confiding. Peafowl too are protected, though for different reasons. The calls of the males echo through the woodlands, the voice of the goddess of learning, the god of war and the god of the occult— all at the same time.

Though primarily a bird reserve, and a fine and spectacular one too, Bharatpur also boasts a few interesting mammals. Blackbuck, one of the most beautiful and rare members of its family, can be seen, and chital or spotted deer roam often in quite sizeable herds. The larger nilgai are also present.

Visiting: The Sanctuary may be reached by road and is thirty miles west of Agra and a hundred miles south of Delhi. There are tracks and roads crossing the area and the staff of the Park can arrange boat trips among the heronries. There is a rest house offering simple accommodation.

Species of particular interest

Blackbuck
  Antilope cervicapra
Chital
  Axis axis
Nilgai
  Boselaphus tragocamelus
Wild Pig
  Sus scrofa
Darter
  Anhinga rufa
Siberian White Crane
  Grus leucogeranus
Sarus Crane
  Grus antigone
Painted Stork
  Ibis leucocephalus
Open-billed Stork
  Anastomus oscitans
Spoonbill
  Platalea leucorodia
Little Egret
  Egretta garzetta
Intermediate Egret
  Egretta intermedia
Bar-headed Goose
  Anser indicus
Comb Duck
  Sarkidiornis melanotos
White-backed Vulture
  Gyps bengalensis
Pondicherry Vulture
  Torgos calvus
Pallas Fishing Eagle
  Haliaetus leucoryphus
Black-shouldered Kite
  Elanus caeruleus

Painted storks (above) breed in their thousands at Bharatpur, a former shooting reserve of a maharajah. But it is not only birds that haunt the swamps. Nilgai and blackbuk can also be found. Nilgai (left) are the more plentiful, the black males contrasting with their brown mates.

Kinabalu, Sabah

Borneo—a land of impenetrable forest, hostile people and immense danger. That, at least, is the general view and it is not a wholly inaccurate one at that. Rugged mountains rise towards the clouds and frequently reach them. Streams rush and fall, churning a way through the otherwise impassable forest that covers every part of the hilly districts. Borneo has some of the last large primitive rain forests left on earth and the whole of the eastern half of the island is covered with rich woodland. Naturally lumber specialists have their business eyes trained upon this great wilderness area and they are unlikely to be put off by tales of head-hunters. Conservation is in its cradle in Borneo and talk of unique ecosystems and invaluable populations of animals carries little weight with those who see forests as future furniture. And yet attempts to preserve the most spectacular area of the country, now known as Sabah, are being made backed by an enlightened but desperately poverty-stricken government.

The traveller in Sabah today would be able to find much interesting wildlife no matter where he went, for much of the forests are the same and game is not concentrated into parks or left as local remnants. Animals are widespread, but nowhere numerous. They are, however, of unique interest and appeal to those whose journeys must include seeing the world's large mammals in their natural habitats.

Mount Kinabalu National Park covers 265 square miles of the highest mountain between the Himalayas and New Guinea—13,455 feet. While the fauna of Borneo is similar to that of neighbouring Java, Sumatra and Malaya the most interesting of the island's species are found on the mountains which have maintained geographical isolation for a long period. Before appetites are whetted it should also be noted that the mountains in this part of the world are permanently shrouded in mist and receive incessant rain throughout the year. The only respite is a short break in the early morning. Over 180 inches of rain per annum fall on these oak forests.

As the visitor progresses upwards through the jungle the wealth of birds everywhere will impinge on his consciousness, but above 8000 feet the Kinabalu mountain rat becomes the most common and ubiquitous animal. Birds are particularly numerous on the unspoilt eastern ridge—the tourist route and Park H.Q. lie on the south-western ridge.

The eastern ridge still holds sambar and barking deer as well as wild pig and cattle and the Sumatran rhinoceros. This animal is now so rare and so widely scattered that its chances of survival must be considered very low. If it is not to become extinct it will be because of its continuing presence in areas like Sabah and National Parks like Kinabalu. The same dreary story could be told of the orang-utan. This old man of the forest was once widespread throughout the rain forests of this part of Asia; today it is progressively confined to fewer and fewer areas of primitive jungle. In Sabah, one of the acknowledged strongholds, the Sepilok Forest Reserve is a stronghold and here the authorities have established a rehabilitation centre where orang-utans can be restored to nature after being captured for the smuggled export market. A few orang-utans manage to exist at Kinabalu—it is certainly one of the most accessible places to see them—though visitors would be better advised to concentrate their search in the Kinabatangan, Segama and Tinkayu river systems which hold half the world population. Among the many myths associated with orang-utans is one that attributes to them the power of speech which they never use for fear of man putting them to work. They are also said to carry off women for wives and on at least one occasion of having begotten a baby man-ape.

Other mammals found at Kinabalu include perhaps the richest primate fauna in the world. Gibbon, leaf monkeys, macaques, tree-shrews in abundance, western tarsier and slow loris all haunt the great trees of the mountain forests. Maroon leaf-monkeys can still be found, but they are good eating and much sought after on that account. There is also a wealth of squirrels ranging from the $2\frac{1}{2}$ feet giant tufted-eared, down to the tiny pygmy squirrels.

The Kinabalu friendly warbler is found here, at nearby Mount Trus Madi, and nowhere else on earth. Other birds like the drongos and tree-pies are more widespread and numerous. As elsewhere in woodland, birds tend to flock together and the forests often seem empty of life. But it is difficult to miss the mountain blackeye and mountain blackbird, both of which are extremely vociferous. In the east, Poring, with its hot springs, provides access to the least disturbed part of the Park. Here the golden-crowned bulbul and white-rumped shama delight the visitor with their songs.

Visiting: Intending visitors should contact the Park authorities on arrival in Sabah. There are few tourist facilities but access can be made through Poring. Leeches and snakes, including one that 'flies', are an ever-present problem.

Kinabalu National Park

SABAH

Species of particular interest

Orang-utan
  Pongo pygmaeus
Maroon Leaf Monkey
  Presbytis rubicundus
Grey Leaf Monkey
  Presbytis spp.
Gibbon
  Hylobates spp.
Pig-tailed Macaque
  Macaca nemestrina
Long-tailed Macaque
  Macaca spp.
Proboscis Monkey
  Nasalis larvatus
Western Tarsier
  Tarsius bancanus
Slow Loris
  Nycticebus coucang
Flying Lemur
  Cynocephalus variegatus
Malayan Pangolin
  Manis javanica
Sumatran Rhinoceros
  Didermocerus sumatrensis
Kinabalu Friendly Warbler
  Bradypterus accentor
Mountain Blackeye
  Chlorocharis emiliae
Mountain Blackbird
  Turdus poliocephalus

Klamath, California, USA

From the Pacific coast the Klamath River runs inland, twisting its way up through the coastal ranges of the mountains of the same name. Higher up it leaves California for Oregon and the Upper Klamath Lake. Here, on both sides of the State Line, lie the Klamath Basin National Wildfowl Refuges at 4000 feet. Surrounded by thick woodland, sagebush and rolling agricultural land, five refuges sit astride the Pacific Flyway and attract perhaps three-quarters of the birds that pass that way. This may be as many as 10 million birds during the fall migration.

Most prominent among them are the geese. Huge numbers of snow, lesser snow and Canada geese blot out the sky as they alight to feed on the crops specially grown for their benefit. A roost of these birds contains literally hundreds of thousands of birds plus even a few white-fronts and Ross's geese. The sheer number of birds makes the area one of the great wildlife paradises. But there are duck here too—millions of them: pintail, American wigeon, canvasbacks, teal, shoveler, blue-winged teal, scaup.

Though all five refuges have these birds, Tule Lake and the Lower Klamath Refuge, between them covering nearly 60,000 acres, are the major spots. They have the greatest concentration of wildfowl in North America.

With the wildfowl come hawks, herons and hosts of shorebirds. There are sandhill cranes in flocks as large as can be found anywhere. And then there are the breeding birds, not as dramatic but an attraction in their own right. Eared, western and pied-billed grebes, the largest colony of white pelicans in the United States at Clear Lake, Caspian terns, great blue herons and American egrets, double-crested cormorants, black-crowned night herons, plus all the usual birds of this part of the American west.

Visiting: There are hotels at Tule Lake, California, and at Klamath Falls, Oregon, which are ideally situated. All the refuges are open free of charge the year round. Contact headquarters at Tule Lake. Plenty of tracks facilitate exploration.

Species of particular interest

Eared Grebe
  Podiceps caspicus
Western Grebe
  Aechmophorus occidentalis
Pied-billed Grebe
  Podilymbus podiceps
Great Blue Heron
  Ardea herodias
American Egret
  Casmerodius albus
Black-crowned Night Heron
  Nycticorax nycticorax
Snow Goose
  Chen caerulescens
Lesser Snow Goose
  Chen hyperborea
Ross's Goose
  Chen rossii
Canada Goose
  Branta canadensis
White-fronted Goose
  Anser albifrons
American Wigeon
  Anas americana
Pintail
  Anas acuta
Blue-winged Teal
  Anas discors
Shoveler
  Anas clypeata
Gadwall
  Anas strepera
Canvasback
  Aythya valisineria
Scaup
  Aythya marila
Sandhill Crane
  Grus canadensis

Descending from the north in their hundreds of thousands, snow geese dominate the autumn and winter scene at Klamath in the western Rockies. These are lesser snow geese, the most numerous of all.

Kruger National Park, South Africa

Covering 8000 square miles the Kruger National Park is at once one of the largest and most visited game parks in the world. In 1968 over a quarter of a million people visited Kruger bringing with them nearly seventy thousand vehicles. And yet the Park lies in the extreme north of the country against the Mozambique border and a considerable distance from the great centres of population. It is an undulating region with vast open grasslands in the south and bush country to the north. Several rivers have cut their way through the landscape creating gorges in some areas and wooded slopes in others. Riverine forests are an important habitat in this basically open area.

Kruger National Park was created in 1926 replacing the Sabie Game Reserve created by President Paul Kruger in 1898. Even at that time the wildlife of South Africa was dwindling fast. The Park was a pathfinder, the first of its kind in Africa. Hunters continued to exploit its game and farmers to encroach upon its boundaries, but slowly more camps for guards were established and gradually the rule of law prevailed.

Of all the Park animals the white rhinoceros is the most notable. At one time this greatest of all the rhinos was incredibly numerous and widespread in the southern third of Africa. A few years after the start of European exploitation of the continent these huge animals were practically extinct and survived only in a small pocket of Zululand at Umfolozi and Hluhluwe. Strict control and protection

Perhaps the ugliest birds in the world, marabou storks (below) are the snappers-up of trifles left unconsidered by the other African scavengers. Fights are frequent and vicious and invariably end in bloodshed. Sable too are plentiful, the fine black males guarding the herds of females (left) as they drink.

has enabled the population to revive once more and the rhinos have been reintroduced into other parts of South Africa including the Kruger. They are extraordinarily tame and photography is almost without risk.

Those other large game-park 'stars' the elephant and hippopotamus are also present in good numbers in Kruger. Buffalo can be found in large herds, and there are no less than seventeen distinct species of antelope. All have at one time or another been found along the banks of the Shingwidzi, a major wildlife concentration point. They include eland, the largest of the antelopes, as well as the beautiful sable and highly localized roan. Kudu, impala and waterbuck can all be found and inevitably attract a variety of predators and scavengers. Lion, leopard, cheetah, spotted hyena, wild dog and jackal all roam the plains of Kruger—the greatest wildlife spectacle outside East Africa.

The rivers are one of the major attractions to the birds of the park as well as to the crocodiles that line their banks. Ground hornbills often walk the park's tracks while the magnificent lilac-breasted rollers perch on the tops of thorns. Ostrich, marabou, guineafowl, francolins, trogons and drongos all add to the medley of life in this superb wildlife area.

Visiting: The Park entrance is 200 miles from Johannesburg, the nearest airport. Over 1000 miles of road runs through the Park and there is a wealth of accommodation to suit all tastes.

Species of particular interest

White Rhinoceros
  Ceratotherium simum
Elephant
  Loxodonta africana
Hippopotamus
  Hippopotamus amphibius
Warthog
  Phacochoerus aethiopicus
Buffalo
  Syncerus caffer
Giraffe
  Giraffa camelopardalis
Wildebeeste
  Connochaetes taurinus
Lion
  Panthera leo
Leopard
  Panthera pardus
Cheetah
  Acinonyx jubatus
Roan
  Hippotragus equinus
Sable
  Hippotragus niger
Impala
  Aepyceros melampus
Eland
  Tamotragus oryx
Steenbuck
  Raphicerus campestris
Wild Dog
  Lycaon pictus
Honey Badger
  Mellivora capensis
Kudu
  Tragelaphus imberbis
Spotted Hyena
  Crocuta crocuta
Jackal
  Canis mesomelas
Vervet Monkey
  Cercopithecus aethiops
Crocodile
  Crocodylus niloticus
Ostrich
  Struthio camellus
Marabou
  Leptoptilos crumeriferus
Secretarybird
  Sagittarius serpentarius
Honey-guide
  Indicator indicator
Lilac-breasted Roller
  Corcacias caudata
Crowned Guineafowl
  Numida meleagris

The square-lipped white rhinoceros, eliminated over much of its range, still finds a stronghold in the great Kruger Park.

Lamington National Park, Australia

Seventy miles south of Brisbane and twenty-five miles from the Queensland coast lies the rugged McPherson Range. Here, covering 50,000 acres, lies the Lamington National Park, rising to a peak of nearly 4000 feet at Mount Wanungra. Though easily accessible and served by a network of forest paths and tracks, Lamington remains an almost unspoilt wilderness. It is served by two famous lodges that are patronized by naturalists from all over the world. At O'Reilly's Guest House, in particular, it is difficult to get away from natural history from dawn to dusk—after that there are usually talks, films and slide shows!

Lamington has a variety of landforms and habitats but the major vegetation is sub-tropical rain forest. There are also eucalypt forests and the mallee (shrubby eucalypt) heath beloved of the scrub turkeys. The close-canopy, jungle-like forests provide a home for a variety of epiphytes, orchids and ferns that attach themselves to the trees, and creepers which wind their way upwards like giant runner-beans. The strangling fig grows in the fork of a tree and, while sending its shoots upwards, also sends roots downwards. Eventually the host tree is strangled by the roots.

Most of the mammals of the park are nocturnal, though around O'Reilly's several species have been encouraged to show themselves. Here are pretty-faced wallabies, red-shouldered pademelon, and various possums, bandicoots and the smaller marsupials. The dingo, Australia's wild dog, is also found here. There are many snakes, some of which will be seen by the careful observer, and other reptiles like the larger lizards.

Over a hundred species of birds have been recorded in the park—not a huge total by any means, but some are exciting

Like a giant kingfisher, the kookaburra (left) or laughing jackass, watches for its reptile or insect prey. Of all the parrots of Australia, none is more colourful than the delightful rainbow lorikeet (right). Gathering in huge flocks at favoured feeding grounds they are generally tame and confiding with people.

Species of particular interest

Pretty-faced Wallaby
  Protemnodon elegans
Red-shouldered Pademelon
  Thylogale thetis
Ringtail Possum
  Pseudocheirops archeri
Short-nosed Bandicoot
  Thylacis obesulus
Short-eared Brush-tail Possum
  Trichosurus caminus
Tiger Cat
  Dasymops maculatus
Marsupial Mouse
  Antechinus spp.
Dingo
  Canis dingo
Eastern Water Dragon
  Physignathus spp.
Wedge-tailed Eagle
  Aquila audax
Grey Goshawk
  Accipiter novaehollandiae
Crimson Rosella
  Platycercus elegans
Wonga Pigeon
  Leucosarcia melanoleuca
Top-knot Pigeon
  Lopholaimus antarcticus
Brush Turkey
  Alectura lathami
Laughing Kookaburra
  Dacelo gigas
Pied Currawong
  Strepera graculina
Paradise Riflebird
  Ptiloris paradiseus
Prince Albert Lyrebird
  Menura alberti
Satin Bowerbird
  Ptilonorhynchus violaceus
Regent Bowerbird
  Sericulus chrysocephalus
Green Catbird
  Ailuroedius crassinostris
Southern Logrunner
  Orthonyx temminckii

and at O'Reilly's there are fine opportunities for photography. Satin and regent bowerbirds perform well and the latter will feed from the hand. The scrub turkey, that avid mound-builder that uses the heat of the sun to create an incubator instead of hatching out its own eggs, is widespread and obvious. In contrast the rufous scrub-bird is so inconspicuous that it was only discovered in 1923. In the gullies Prince Albert's lyrebird will be found and there are grey goshawks and the fast-declining wedge-tailed eagles to be seen.

Visiting: On the way from Brisbane stop at the Currumbin Bird Sanctuary on the Gold Coast where every morning and afternoon tens of thousands of lorikeets come to be fed. Continue southwards to the Park and stay at Binna Burra, where tame birds include the kookaburra, or continue to O'Reilly's at Green Mountains.

Lapland, Sweden, Finland and Norway

In northernmost Scandinavia lies Europe's last remaining wilderness—Lapland. It is no respecter of boundaries and though we refer to Swedish Lapland and to Finnish Lapland, the whole area, including parts of Norway, is best treated as one unit.

Lapland is the land of the Lapps and their herds of reindeer. Domesticated for hundreds of years, it is still as well to remember that reindeer are the same species as the caribou of North America. Those who know only that continent will have an idea of what caribou country is like. Lapland lies at the northern edge of the great coniferous forests that cover so much of Scandinavia and northern Europe—at the temperate tree-line. To the north stretch the great open and tree-less tundra, but the most interesting of Lapland's habitats is the taiga, the transitional zone.

Because of this 'edge-effect' Lapland boasts a unique (in Europe) series of habitats and as a result accommodates a large range of animals. Frozen for most of the year it comes alive in May and June. Some say that it becomes too alive—there are mosquitoes everywhere and only netting and liberal use of repellents enables the visitor to survive. The Scandinavians, however, are great out-of-doors people and they have taken all of the precautions they can to make accommodation both comfortable and pest-free. Mosquitoes need water to breed and that is exactly what Lapland has got. Lakes, pools and marshes cover the landscape and create incredible richness.

Much of the wildlife of Lapland has been destroyed—it is a wilderness that has been inhabited for hundreds of years—but there are still a few large mammals to be seen, though not on the North American scale. Reindeer are predominantly domesticated, but there are a few herds of feral animals that are missed at round-up and which can be regarded as the genuine article. Arctic foxes are sometimes encountered and blue hares are reasonably common. But in the main the mammals are small—red squirrel, martens and lemming. The occasional brown bear can be seen as can the even rarer lynx and wolf. But all three predators have been virtually shot out of existence.

If mammals are scarce, birds are incredibly plentiful. Most are migrants that come simply for the summer, but others like the owls and woodland species manage to survive through the winter.

Birds are abundant throughout Lapland,

indeed every little pool seems to have its own pair of red-throated divers, sandpipers and skuas, but inevitably there are concentration points. Lake Inari, a huge lake dotted with islands, holds hordes of duck. The woodlands at Enontekio are alive with small birds, as are those at Abisko in Sweden which lacks a road but boasts a railway. Karigasniemi is well out on the tundra and alive with waders, skuas and other open-land species. Here one can find whooper swans breeding, with long-tailed duck, rough-legged buzzards, hundreds of sandpipers, including the elusive broad-billed sandpiper that looks like a snipe. Phalaropes and long-tailed skuas are everywhere and shorelarks nest along with Lapland and snow buntings. This is also a possible area for breeding bar-tailed godwits.

To the south the woodlands have great grey, Tengmalm's and other boreal owls. Siberian and crested tits flit through the tree-tops with Siberian jays, crossbills and bramblings. While the woodland marshes are the haunt of the vocal and dancing common cranes.

Those who enjoy the wide open spaces of Lapland with the fast, but dirt-surfaced roads, will be drawn further and further north towards the harsh and cold area of the Varanger Fjord—the only accessible part of the Arctic mainland. Here truly Arctic species like Brunnich's guillemot and Stellar's eider can be found along with human enterprise in the form of smelly fish processing plants and canning factories. But it's a wild landscape and full of birds.

Visiting: There is an excellent network of dirt roads and a good variety of accommodation in typically Scandinavian lodges. There are well-marked tracks and paths and excellent maps to aid the walker. National Parks and Reserves are numerous and information is available on each.

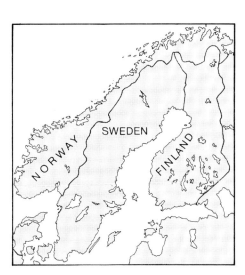

The swamps and marshes of Lapland are a home to hordes of migratory birds including the large, but elusive crane.

Species of particular interest

Reindeer
  Rangifer tarandus
Brown Bear
  Ursus arctos
Lynx
  Felis lynx
Wolf
  Canis lupus
Red Squirrel
  Sciurus vulgaris

A common enough sight in Lapland, a herd of reindeer (below left) cross a forest track. The call of the dancing cranes (below) is far more elusive, while a sight of a lynx (above left) is reserved for the extremely fortunate.

Red-throated Diver
  Gavia stellata
Steller's Eider
  Polysticta stelleri
Whooper Swan
  Cygnus cygnus
Rough-legged Buzzard
  Buteo lagopus
Spotted Redshank
  Tringa erythropus
Greenshank
  Tringa nebularia

Broad-billed Sandpiper
  Limicola falcinellus
Temminck's Stint
  Calidris temminckii
Little Stint
  Calidris minuta
Long-tailed Skua
  Stercorarius longicaudus
Arctic Tern
  Sterna paradisea
Tengmalm's Owl
  Uria lomvia

Brunnich's Guillemot
  Aegolius funereus
Lapland Bunting
  Calcarius lapponicus
Snow Bunting
  Plectrophenax nivalis

Mallacoota Inlet National Park, Australia

Two lakes created by the drowning of a river valley and now consisting of inter-connected waters with an outlet to the sea at the Bass Strait. Mallacoota lies at the far eastern corner of Victoria near the New South Wales border. The National Park covers over 11,000 acres of land adjoining the lakes and is a mixture of water and woodland beautifully set against a backdrop of distant mountains. The lakes are fed by the Genoa River and the exit to the sea is shallow and narrow. Skill and knowledge of local conditions are necessary to navigate it.

Mallacoota is a national park in the Yellowstone mould—an area of such beauty that pioneers felt bound to try to protect it. Mallacoota survived a mini gold rush (the remains can be seen at the upper reaches of Bottom Lake) but is now a fast-developing tourist resort. Fishing, log cabins, pleasure boats, golf, surfing, these are the new leisure activities that threaten to ruin the area as the refuge it undoubtedly is.

The park consists of forests of eucalyptus with thickets of banksia, and a few rainforest gullies. To the north lies the Howe Range with its epiphytic ferns and orchids, while to the south, across the Betka River, lie heathlands that provide a home for the rare ground parrot.

The twenty resident species of mammals are all difficult to find, shy creatures that emerge only at night. Fortunately the largest, the grey kangaroo, can be seen early in the morning and in the growing dusk on the golf course and at Gipsy Point well away from the coast. Gliding possums can be found here and there—the yellow-bellied glider and the sugar glider. Ringtail and brush-tail possums frequent the tea-tree scrub near the water's edge and even come to picnic tables. There is a rookery of fur seals down the coast at Wingan National Park which can be easily visited.

Not surprizingly Mallacoota is a superb place for birds, with a list of over 200 species. The islands and foreshore hold colonies of crested terns, silver and Pacific gulls, the cosmopolitan Caspian terns, and several plovers. There are pelicans and swamp harriers and all five species of Australian cormorants. In the forests lyrebirds can entrance those fortunate enough to see their extra-ordinary courtship displays. And satin bowerbirds, king parrots and crimson rosellas can all be seen. The heathland to the south has honeyeaters of its own as well as the ground parrot and brown quail. The open forest and banksia scrub is the home of huge flocks of rainbow, musk and little lorikeets that sometimes make other habitats seem empty in comparison. Above it all fly the white-breasted sea eagles, the masters of beautiful Mallacoota.

Visiting: There are hotels, motels and camp sites at Mallacoota township and access to the Park is unrestricted. Boats can be hired.

Species of particular interest

Grey Kangaroo
  Macropus canguru
Yellow-bellied Glider
  Petaurus australis
Sugar Glider
  Petaurus breviceps
Ringtail Possum
  Pseudocheirops archeri
Brush-tail Possum
  Trichosurus vulpecula
Swainson's Marsupial Mouse
  Antechinus swainsoni
Stuart's Marsupial Mouse
  Antechinus stuartii
Fur Seal
  Arctocephalus doriferus
Australian Pelican
  Pelecanus conspicillatus
Oystercatcher
  Haematopus ostralegus
White Egret
  Egretta alba
Reef Heron
  Egretta sacra
Royal Spoonbill
  Platalea regia
Yellow Spoonbill
  Platalea flavipes
White-breasted Sea Eagle
  Haliaetus leucogaster
Swamp Harrier
  Circus approximans
Crested Tern
  Sterna bergii
Lyrebird
  Menura novaehollandiae
King Parrot
  Alisterus scapularis
Ground Parrot
  Pezoporus wallicus
Rainbow Lorikeet
  Trichoglossus haematodus
Satin Bowerbird
  Ptilonorhynchus violaceus

AUSTRALIA

Mallacoota
Inlet National Park

Mallacoota Inlet is a bird paradise, but grey kangaroos (bottom) can often be found along its sheltered southern shores. Below, crested terns nest along the seashore in vast numbers and are often accompanied, as here, by the red-billed silver gulls.

Lake Manyara National Park, Tanzania

Though only 123 square miles in extent, of which 88 square miles are lake, Lake Manyara National Park boasts a wide variety of country. Situated anywhere else but along the usual direct approach to Ngorongoro and the Serengeti Plains, it would be the gem of the collection. As it is, many tourists rush by without a second thought for this superb wildlife area. Even naturalists tend to dismiss it as a good place for elephants and one where lions sleep in trees like leopards.

Lake Manyara National Park lies seventy-five miles south-west of Arusha and includes an area bordering the north-western corner of the lake. Its boundaries encompass the western escarpment of the Great Rift Valley that runs north to south and which, at this point, has no eastern wall. A series of rivers flow down the escarpment to feed the lake, those in the north being particularly interesting for their origin in springs at the porous base of the rift escarpment.

Though only twenty-five miles long there are many side tracks that enable a full exploration to be made. Almost all visitors will manage to see the largest and most impressive of Manyara's animals— lion, buffalo, elephant, zebra and giraffe. But there is also a good chance of black rhinoceros and leopard, particularly in the early morning or in the evening.

Every part of Manyara is of interest, the ground water forest in the north with its baboons, the open swamp with its herd of four hundred buffalo, the Ndala River with its elephant research camp, and the acacia woodland with its climbing lions. Incidentally the theory is that the lions climb trees to avoid the troublesome tsetse flies—though the intending visitor should note that these do bite but do not carry sleeping sickness.

Most of the usual East African large mammals can be found within the park, including hippopotamus, dik-dik, reed-buck and bushbuck. Birds too are abundant and varied. At times lesser flamingoes descend in their thousands. White-backed duck and the tiny pygmy goose can be seen with all the usual storks, ibises, herons and egrets. More than thirty raptors have been noted including palm nut vulture and crowned hawk eagle. Many Eurasian waders haunt the shores of Lake Manyara, sometimes with the rare chestnut banded sand plover and the umbrella-simulating black heron.

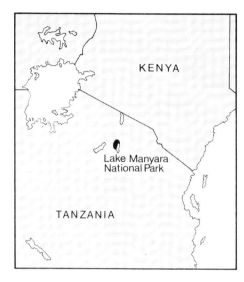

Visiting: There is only one entrance to Manyara. It forks southwards off the main Arusha–Seronera road as it begins to climb westwards up the Rift Valley escarpment. The Park Office and camp site are also here. The magnificently sited Lake Manyara Hotel offers superb views over the park and the lake from a spur of the escarpment.

Species of particular interest

Lion
  Panthera leo
Leopard
  Panthera pardus
Elephant
  Loxodonta africana
Buffalo
  Syncerus caffer
Black Rhinoceros
  Diceros bicornis
Hippopotamus
  Hippopotamus amphibius
Giraffe
  Giraffa camelopardalis
Klipspringer
  Oreotragus oreotragus
Silver-backed Jackal
  Canis mesomelas
Bushbuck
  Tragelaphus scriptus
Reedbuck
  Redunca redunca
Goliath Heron
  Ardea goliath
Black Heron
  Melanophoyx ardesiaca
Dwarf Bittern
  Ardeirallus sturmii
Hadada Ibis
  Hagedashia hagedash
Lesser Flamingo
  Phoeniconais minor
White-backed Duck
  Thalassornis leuconotus
Pygmy Goose
  Nettapus auritus
Cuckoo Falcon
  Aviceda cuculoides
Palm-nut Vulture
  Gypohierax angolensis
Chestnut-banded Sand Plover
  Charadrius venustus
Terek Sandpiper
  Xenus cinereus

Lake Manyara is best known for its tree-climbing lions (top) that apparently spend most of their lives asleep. Giraffes (bottom) cannot afford such luxuries and are ever watchful for predators, particularly when they have young.

Mount McKinley National Park, Alaska, USA

At 20,320 feet Mount McKinley is the highest point on the North American continent and the centre of this 3000 square mile National Park. Most of the Park lies above the tree line, hereabouts some 3000 feet, though wildlife takes little notice of boundaries. While the valleys are coated with a thick layer of spruce the Park itself consists of Alpine tundra with stands of rhododendron and dwarf bushes. Higher the land is permanently covered with ice and snow. Many of the most beautiful areas boast lakes and ponds and it is the waterbirds that first attract the attention. Great northern divers, horned (Slavonian in Britain) grebes, scaup and Barrow's goldeneye engage their varied and often noisy courtship on the lakes, while harlequin duck cascade down waterfalls and rapids with incredible skill.

Tundra birds are as confiding here as anywhere, and rock and white-tailed ptarmigans sit tight as the visitor walks by without noticing. Golden eagles soar overhead while the thick woodlands hide the hawk owl and gray jay.

Though birds are so obvious, there are many attractive and large mammals in the Alaska Range. Grizzly bear are numerous along the valleys and can be seen at the appropriate time of the year gorging themselves on the run of sockeye salmon in the shallow rivers. Caribou too are numerous during the spring and autumn migration periods and, even as far south as this, packs of wolves follow the herds. Moose are seldom seen in large numbers but individuals are widespread among the valley swamps. Beavers are also valley animals and here, at least, they have not been shot out of existence.

Higher up the slopes the shrill whistles of the hoary marmot echo round the crags while on the tops themselves the all-white dall sheep step gracefully from rock to rock. Snowshoe hares are quite common inhabitants of the higher slopes coming lower in winter when snow covers the ground. Their well furred feet enable them to scamper across soft snow at quite remarkable speeds.

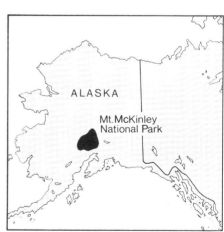

Visiting: McKinley can be reached by car from Anchorage and has only one road. The McKinley Hotel offers accommodation through the summer, and there are seven camp sites. The hotel has its own airstrip and has cars for hire. The Park can also be reached by railway.

The richest wildlife area in Alaska, Mount McKinley is the home of large flocks of dall sheep (left) that climb the steepest rock faces. In the woods below the brown bear (right) ekes out a living from whatever comes his way.

Species of particular interest

Grizzly Bear
  Ursus arctos
Wolf
  Canis lupus
Barren-ground Caribou
  Rangifer tarandus
Moose
  Alces alces
Red Fox
  Vulpes vulpes
Wolverine
  Gulo gulo
Dall Sheep
  Ovis dalli
Hoary Marmot
  Marmota spp.
Collared Pika
  Ochotona collaris
Porcupine
  Erethizon dorsatum
Arctic Ground Squirrel
  Citellus spp.
Red Squirrel
  Tamiasciurus spp.
Beaver
  Castor fiber
Lynx
  Felis lynx
Snowshoe Hare
  Lepus americanus
Great Northern Diver
  Gavia immer
Horned Grebe

Podiceps auritus
Greater Scaup
  Aythya marila
Barrow's Goldeneye
  Bucephala islandica
Harlequin Duck
  Histrionicus histrionicus
Golden Eagle
  Aquila chrysaetos
Surfbird
  Aphriza virgata
Wandering Tattler
  Heteroscelus incanum
Hawk Owl
  Surnia ulula
Rock Ptarmigan
  Lagopus mutus
White-tailed Ptarmigan
  Lagopus leucurus
Spruce Grouse
  Canachites canadensis
Gray Jay
  Perisoreus canadensis

Inhabitants of the forest zone, hawk owls (left) and porcupines (below) utilize trees in different ways. For the owl a conifer provides a look-out post and possible nest site; for the porcupine a sallow is a meal in itself. Both are frequently seen at McKinley.

Merja Zerga, Morocco

Merja Zerga is a shallow inter-tidal lagoon on the Atlantic coast of Morocco, north of Casablanca. It is surrounded by a low-lying region known as the Rharb which is flooded every winter and is then the resort of thousands of wildfowl and waders that move southwards from Europe and western Asia. There are few wildlife areas of great merit in north Africa or in the Arab world as a whole, but Merja Zerga is so full of birds that few could fail to be impressed.

No more than six miles square it empties almost completely at low tide, when the mud and sand-banks provide food for a host of migrants and winter visitors. Because it has only a narrow opening to the sea, salt water seldom penetrates very far and the regular daily flood is predominantly fresh. This enables large numbers of fresh water waders like wood sandpipers and black-tailed godwits to find a living, along with masses of cattle and little egrets.

Flamingoes, passing along the coast from their breeding grounds in Spain and France, regularly stop-over, though these birds do breed in southern Morocco in an inaccessible area along the River Dra. Masses of terns consist predominantly of migrant black terns, dipping delicately to feed from the water's surface though, in autumn at least, the large lesser crested and royal terns are present as well.

Both species, being basically tropical in origin, are seldom found further north. Hosts of small waders, called 'peep' in America, are dominated by dunlin, though little stints are present in numbers that would astound the average European enthusiast. Ruff call a halt here in large numbers prior to their long trans-Saharan migration.

In winter Merja Zerga is one of the most important areas for waders along the eastern Atlantic seaboard and may be second only to the great Waddensea along the German and Dutch coasts. It also boasts a number of non-aquatic birds that are of interest including the African marsh owl that is found no nearer Europe. The lagoon is flanked by a broad belt of eucalyptus along the dune-lined coast that gives shelter to a number of smaller birds.

Of all the attractions of this area coot and slender-billed curlew seem to be poles apart. But when birds are present in really large numbers, say 100,000, it's difficult not to be impressed, even if they are coot! In contrast the slender-billed curlew is one of the rarest and most elusive of birds. It breeds in southern Russia and then migrates eastwards through the Mediterranean to winter in Tunisia and Morocco. As many as 300 have been noted at Merja.

Visiting: Highly accessible, most of its surface can be seen from the surrounding roads, and the seaside resort of Moulya Bou-Selem lies at its mouth. There is a first class European hotel at Moulya Bou-Selem, and the whole area can be reached by passable roads with the help of a navigator and a good map. Local boats can penetrate along the shallow major channels at high tide and are cheap to hire.

A curious mixture of northern and southern influences is apparent at Merja Zerga. From the north come vast hordes of migratory waders like these black-tailed godwits (below). While of more southern origins are the African marsh owls (left) which quarter the marshes in search of prey.

Merja Zerga

MOROCCO

Species of particular interest

Flamingo
   Phoenicopterus ruber
Cattle Egret
   Ardeola ibis
Little Egret
   Egretta garzetta
Lesser Crested Tern
   Sterna bengalensis
Royal Tern
   Sterna maxima

Slender-billed Curlew
   Numenius tennirostris
Black-tailed Godwit
   Limosa limosa
Black Tern
   Chlidonias niger
Whiskered Tern
   Chlidonias hybrida
Caspian Tern
   Hydroprogne tschegrava
Little Stint
   Calidris minuta

Myvatn, Iceland

Set in surroundings reminiscent of the moon, a land where astronauts receive their pre-moonshot training, Myvatn is a true oasis of life in the midst of a desert. Rocky lava flows, volcanic craters, cones of ash, geysers and sulphur pools make the surroundings of this huge lake both barren and ugly. But at these latitudes the combination of fresh shallow water and long hours of sunlight inevitably create immense richness.

Myvatn actually means 'lake of flies' and that is just what it is in mid-summer. Midges, uncountable billions of midges, hatch from the surface and turn the air a buzzing grey. They are not mosquitoes and they do not bite, but they do drive people mad. They also drive the fish mad. Trout by the thousand spend all their time guzzling themselves, feeding on this rich food supply. In turn this has given rise to a host of fish eating predators in the form of thousands of duck. It all starts in the incredible bloom of vegetation that greets the summer. Vegetation, microscopic life, midges, fish, duck, all are interdependent. The food chain ends with the huge gyrfalcon, the king of Myvatn.

Though it is only five miles across, Myvatn has an incredibly indented coastline. Peninsulars continually cut off open views and create basins and lagoons. It has been called the bird-watcher's Mecca simply because there is nowhere else in Europe where so many duck can be seen, but even on the world scale the birds of Myvatn are noteworthy.

Four species are of American origin and can be found nowhere else on the eastern side of the Atlantic. The great northern diver breeds here and migrates south-eastwards to winter off the western coasts of Britain and Ireland. But harlequin duck, Barrow's goldeneye and American wigeon venture no farther eastwards, forming a unique attraction to the European visitor. The harlequins live on the Laxa River and plunge intrepidly through its tumbling waters.

A total of fifteen species of duck breed at Myvatn with perhaps a population of 10,000 pairs. In late summer when they have bred, the number of birds on the lake is several times this figure. But there are other birds too. Slavonian grebe, Arctic tern, the dainty red-necked phalarope, whooper swan—all can be found among the duck of Myvatn. There is a lake where scaup is the most common bird every day. Only one species of mammal is of any note—the mink, which has been introduced and is now a serious predator of duck.

ICELAND

Myvatn

Visiting: Can be reached daily by road and air from Reykjavik. There are two hotels and no restriction on access. Visitors should see the sulphur pools at Namafjall and be sure not to disturb the nesting gyrfalcons should they find them. The government imposes heavy fines on anyone found guilty of disturbing these birds.

Species of particular interest

Mink
  Mustela vison
Great Northern Diver
  Gavia immer
American Wigeon
  Anas americana
Harlequin Duck
  Histrionicus histrionicus
Barrow's Goldeneye
  Bucephala islandica
Scaup
  Aythya marila
Whooper Swan
  Cygnus cygnus
Wigeon
  Anas penelope
Pintail
  Anas acuta
Shoveler
  Anas clypeata
Long-tailed Duck
  Clangula hyemalis
Goosander
  Mergus merganser
Gyrfalcon
  Falco rusticolus
Red-necked Phalarope
  Phalaropus lobatus
Arctic Tern
  Sterna paradisaea

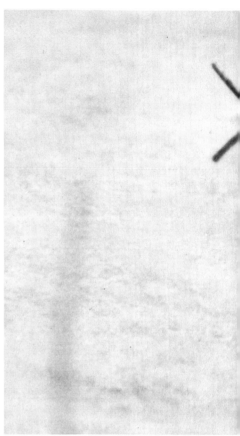

The volcanic landscape of Myvatn (bottom) has produced a jagged coastline with sheltered bays and islands providing ideal breeding grounds for the northern waterbirds. Red-necked phalaropes (top) are a common sight as they sit buoyantly on the water's surface.

Nairobi National Park, Kenya

Nowhere on earth is it possible to be a transit passenger at the airport and in the space of an hour or two see so many wild animals of such a variety of species. Nairobi National Park covers a mere forty-four square miles and lies less than five miles from the city centre but here lives a community of African plains game, their predators and scavengers in what seems almost a European safari park There is only one real difference—in spite of being surrounded by cars and camera-toting tourists, these animals are wild, free and dangerous.

Many of the larger game animals can be seen easily. Lion and leopard are reasonably common, and cheetah seem to be more frequently photographed here than anywhere else. Zebra can be found alongside many of the smaller gazelles, and Masai giraffe frequent the acacia steppes.

The variety of habitats, from plain to forest, with hills and broken land, river valleys and dams gives rise to a long list of birds, though as a matter of fact birds do not always appear numerous in the park. The augur buzzard is, perhaps, the most common raptor, though all six East African vultures can be found along with several eagles including both crowned and martial. The plains birds include guineafowl, bustards, ostrich and crowned crane. Water birds have increased since the construction of the dams and now include the attractive black crake, several storks and ibises and a wealth of duck and geese. Eurasian waders and harriers can be numerous at the appropriate seasons.

Because of its situation, week-ends often find the park crowded and difficult to work with safari vans everywhere. Even week-days can be busy. For this reason enthusiasts usually make a point of an early start when there is always the chance of seeing some of the more secretive nocturnal species as well as avoiding the crowds.

Visiting: The main gate is off the Langata road from Nairobi and the Park is criss-crossed by a network of roads and tracks which must be kept to. Many spots are noted for particular animals, and visitors are advised to buy a map and seek advice from the guards on entering. There is no accommodation within the Park.

Though within a stone's throw of Nairobi's city centre, cheetahs are among the fine collection of wildlife found within the confines of the National Park.

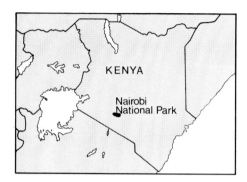

Species of particular interest

Cheetah
  Acinonyx jubatus
Lion
  Panthera leo
Leopard
  Panthera pardus
Burchell's Zebra
  Equus burchelli
Hippopotamus
  Hippopotamus amphibius
Masai Giraffe
  Giraffa camelopardalis
Coke's Hartebeest
  Alcelaphus buselaphus
Wildebeeste
  Connochaetes taurinus
Waterbuck
  Kobus ellipsiprymnus
Impala
  Aepyceros melampus
Thomson's Gazelle
  Gazella thomsoni
Grant's Gazelle
  Gazella granti
Eland
  Taurotragus oryx
Ostrich
  Struthio camellus
Black-headed Heron
  Ardea melanocephala
Marabou Stork
  Leptoptilos crumeniferus
Yellow-billed Stork
  Ibis ibis
Sacred Ibis
  Threskiornis aethiopicus
Secretarybird
  Sagittarius serpentarius
Martial Eagle
  Polemaetus bellicosus
Crowned Eagle
  Stephanoaetus coronatus
Bateleur
  Terathopius ecaudatus
Pale Chanting Goshawk
  Melierax poliopterus
Helmeted Guineafowl
  Numida mitrata

Two buck Grant's gazelles (above right)
face up during the annual rut. They are
among the most common of the animals of
Nairobi National Park. Less numerous, but
still often seen, are the gigantic eland (right)
and statuesque ostrich (above).

Lake Naivasha, Kenya

Only fifty miles north-west of Nairobi lies Lake Naivasha, one of the very few fresh-water lakes in Kenya. Though there is no apparent outlet from the lake it is generally assumed that water flows away beneath the ground thus keeping it from becoming a soda lake like its better known neighbour Nakuru. The fresh water gives Naivasha a character all its own that is instantly recognisable. The shallow water supports an incredible carpet of large pink water lilies among which can be found one of the greatest collections of waterbirds in the world. Dead trees stand grotesquely in the shallows, providing perches for a great variety of herons and egrets. The water level of Naivasha does not fluctuate seasonally but varies irregularly over long periods of time. Thus trees can survive, even though for part of their life their roots are submerged.

Though Nakuru is the place for seeing flamingoes, it is otherwise rather thin on species, whereas almost every East African waterbird can be seen at Naivasha. Black crakes creep through the jungle of water-lilies and floating islands of papyrus while jacanas use their over-extravagant feet to spread their weight over the lily pads that give them their alternate name of lily-trotter. Gallinules stand in small groups while crested coots scurry away over the water's surface when disturbed.

But more than anything else at Naivasha it is the group of birds that go under the awkward name of 'long-legged

The lily-strewn waters of Lake Naivasha are alive with birds. Small parties of hottentot teal (left) hide among the pink flowers, while the dainty jacana (below) builds its floating nest among them.

wading birds' that grabs the attention.

Vast numbers of birds are everywhere, all feeding on the wealth of fish that the rich lake waters support. Grey herons, cattle and little egrets and squacco herons are abundant. But discerning visitors, especially those who are familiar with the fauna of southern Europe, will find their attention drawn to black-headed herons and great white egrets, to yellow-billed egrets and the superb goliath herons, while the very fortunate may encounter the rare black heron. This bird hunts beneath an umbrella that it forms by folding forward its wings over its head.

Storks and ibises are there in plenty with the yellow-billed stork—frequently, and misleadingly, called the wood ibis—standing in attractive off-duty groups around the shallows. Sacred ibises look austere in their white plumage and hadadas somewhat strange and primeval in their irridescent greens. Duck by the thousands haul out on the banks and islets, and European waders find the lush conditions suitable for their winter stay. Overhead the Eurasian harriers quarter the beds to surprise some duck or rail, an increase in competition for the resident African marsh harrier. The African fish eagles fear competition from no one, though ospreys are more common in winter. Each pair of eagles stakes out its territory, which must of necessity include a nest site and a fishing beat. Thus the shore of the lake is divided up into short sections each owned by a different pair of fish eagles. Young birds are beaten off by the established pairs and spend their time searching for carrion and fishing in the no-man's land in the centre of Naivasha.

Nearby is Hell's Gate, a gorge with a reputation for producing birds of prey. In particular most visitors are rewarded with excellent views of lammergeiers as they sail past at close range on the updraughts of air along the rim of the gorge. Verreauex's eagle can also be found.

Mammals are not a speciality of Naivasha though the early rising bird-watcher is likely to see a gazelle or two. The sanctuary of Crescent Island provides a home for a few larger mammals and Hell's Gate is good for rock dwelling species like hyrax and klipspringer.

Visiting: Naivasha is one-and-a-half hour's drive from Nairobi and close to the road. Boats may be hired at the lakeside villages and at the Naivasha Lake Lodge, whose cabins skirt the edge of this paradise. Hell's Gate lies nine miles to the south-east.

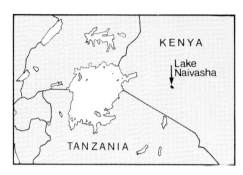

Two characteristic birds of Naivasha. Sacred ibises (bottom) find the rich waters alive with frogs and crayfish, while the pink-backed pelican (top) feeds on the smaller fish in which the lake abounds. Both species can be regularly seen, though generally in small numbers.

Species of particular interest

Klipspringer
  Oreotragus oreotragus
Pink-backed Pelican
  Pelecanus rufescens
Goliath Heron
  Ardea goliath
Cattle Egret
  Bubulcus ibis
Squacco Heron
  Ardeola ralloides
Night Heron
  Nycticorax nycticorax
Hammerkop
  Scopus umbretta
Open-billed Stork
  Anastomus lamelligerus
Marabou Stork
  Leptoptilos crumeniferus
Yellow-billed Stork
  Ibis ibis
Sacred Ibis
  Threskiornis aethiopicus
Glossy Ibis
  Plegadis falcinellus
African Spoonbill
  Platalea alba
Hottentot Teal
  Anas punctata
Fulvous Tree Duck
  Dendrocygna bicolor
Egyptian Goose
  Alopechen aegyptiacus
Ruppell's Vulture
  Gyps ruppellii
Black Crake
  Limnocorax flavirostra
Crested Coot
  Fulica cristata
Crowned Crane
  Balearica regulorum
African Jacana
  Actophilornis africanus
Blacksmith Plover
  Hoplopterus armatus
Painted Snipe
  Rostratula benghalensis
African Skimmer
  Rynchops flavirostris
Malachite Kingfisher
  Corythornis cristata

Lake Nakuru National Park, Kenya

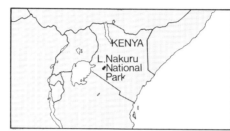

Nakuru and flamingo—the words are almost synonomous and certainly associated in the minds of naturalists. It was here that Roger Tory Peterson described the gathering of flamingoes as the greatest bird sight on earth, yet Lake Nakuru National Park is considerably more than that. The Park lies immediately south of the town of Nakuru which is ninety-seven miles north-west of Nairobi and connected to the capital by a fast modern road. Extending only to twenty-four square miles and, therefore, by East African standards, tiny, most of the Park consists of the lake itself. But the surrounding land is not devoid of wildlife. A herd of hippopotami can be found among the marshes in the north-east, and zebra, giraffe, impala and the commoner gazelles are all usually on view. Bohor reedbuck and bushbuck are difficult to find but definitely present.

Lake Nakuru is an alkaline lake that attracts many water birds apart from the flamingoes. Sometimes over a million of these long-legged and long-necked pink fantasies gather in a tight mass along the shoreline. But they are temperamental birds throughout their range and cannot be relied upon to present the appropriate spectacle. The visitor coming from Nairobi would be well advised to cast an eye over Lake Elmenteita between Naivasha and Nakuru along the way and, if unsuccessful at Nakuru, continue northwards to Lake Hannington. This off-the-beaten-track area is not only a regular flamingo haunt but is also the site of many hot springs—here you can get flamingoes and steam baths in the same picture.

Most of Nakuru's flamingoes are lesser flamingoes, but there are always numbers of greater flamingoes among the flocks. Strangely enough the two species feed side by side and by the same method, but take different food from the water. The birds do not breed at Nakuru but take off for the huge and remote Lake Natron across the Tanzanian border. Some four hundred different birds have been recorded for the Nakuru area. As one would expect this includes many aquatic species but the diversity of the surroundings is such that representatives of almost every likely avian group can be found. While the resident plovers have the shore to themselves between April and September, for the rest of the year they are joined by a host of Eurasian waders. Curlew, wood and green sandpipers trips along together with little stints, marsh sandpipers, ruff and greenshank. Overhead the Eurasian marsh terns, whiskered and white-winged black, vie for attention with the gulls.

Birds of prey are numerous too with fish eagles posted at regular intervals along the shore and Verreaux's eagles nesting at Baboon Rocks to the west.

Among the gems of the collection are the colourful starlings, shrikes, rollers and sunbirds. No less than seven species of sunbirds can be seen at Nakuru plus another two that are irregular visitors.

Visiting: The Park lies to the south of Nakuru and is easy to reach from there. There is plenty of accommodation in Nakuru and the Park has its own camp site.

Species of particular interest

Bohor Reedbuck
  Redunca redunca
Hippopotamus
  Hippopotamus amphibius
Rock Hyrax
  Procavia capensis
Masai Giraffe
  Giraffa camelopardalis
Waterbuck
  Kobus ellipsiprymnus
Impala
  Aepyceros melampus
Greater Flamingo
  Phoenicopterus ruber
Lesser Flamingo
  Phoeniconais minor
Cape Wigeon
  Anas capensis
Maccoa Duck
  Oxyura maccoa
Hammerkop
  Scopus umbretta
White Pelican
  Pelecanus onocrotalus
Darter
  Anhinga rufa
Long-crested Hawk Eagle
  Lophoaetus occipitalis
Grey-headed Kingfisher
  Halcyon leucocephala
Little Bee-eater
  Melittophagus pusillus

A sight which has made Lake Nakuru the best known bird place on earth and has turned ornithologists into poets. A flight of lesser flamingoes takes off from the lake's shallow waters.

Neusiedler See, Austria

Though it lies on the great Hungarian plains Neusiedler See, or Lake Neusiedl, as it is often called, is part of Austria. It is a shallow steppe lake, some fifteen miles long fringed with some of the largest reed beds in Europe. It is not immediately apparent that this is one of the greatest of European wetlands and bird sites. Almost wherever you stand around Neusiedl you can see reeds and in summer the beds are alive with LBJs (not Lyndon Baines Johnsons but Little Brown Jobs). Most ornithologists would make some sort of claim to be able to identify at least a proportion of these birds, but they are very similar, and the less passionate about small warblers would be well advised to give them a miss and concentrate instead on the larger fare of herons and raptors.

Spoonbills feed in the shallows along with purple and night herons. The rare great white egret breeds no further to the west and is common in Europe only in the Danube Delta of Romania. Its long breeding plumes made it one of the favourites in the days when no lady was properly dressed unless she was well feathered. Bitterns spring from the reeds and perform the occasional lap of honour, though for most of their time a boom from the depths is all that can be observed of their presence. Little bitterns, and the pink and black-backed males in particular, are more appealing though they do not fly more than a few yards before disappearing.

As at almost every other Eastern European wetland the greylag goose and white stork are often very much in evidence. The former, the predecessor of the domestic goose and those paté de fois gras cruelties, is common. The stork, the symbol of fertility, is encouraged and nests on the chimneys of village houses.

Most of the birds of prey are rare visitors though enthusiasts do get a great deal from Tadten where views across the Hungarian border can be obtained and where spotted, lesser spotted and imperial eagles can be seen. Marsh and Montagu's harriers quarter the reed beds and red-footed falcons perch on the telegraph wires or hang on the wind like gregarious kestrels.

Bearded tits fly over the tips of the reeds like so many miniature pheasants and penduline tits construct their elaborate nests hanging from a bough along a dyke or ditch. But of all the small birds none attracts the dedicated like the river warbler.

Visiting: Various villages round the lake provide accommodation and there are several camp sites.

Species of particular interest

Red-necked Grebe
  Podiceps grisegena
Greylag Goose
  Anser anser
Ferruginous Duck
  Aythya nycora
Spoonbill
  Platalea leucorodia
Great White Egret
  Egretta alba
Purple Heron
  Ardea purpurea
Bittern
  Botaurus stellaris
Little Bittern
  Ixobrychus minutus
White Stork
  Ciconia ciconia
Black Stork
  Ciconia nigra
Imperial Eagle
  Aquila heliaca
Spotted Eagle
  Aquila clanga
Lesser Spotted Eagle
  Aquila pomarina
Marsh Harrier
  Circus aeruginosus
Montagu's Harrier
  Circus pygargus
Red-footed Falcon
  Falco vespertinus
Kentish Plover
  Charadrius alexandrinus
Black-tailed Godwit
  Limosa limosa
Avocet
  Recurvirostra avosetta
Roller
  Coracias garrulus
Lesser Grey Shrike
  Lanius minor
Bluethroat
  Luscinia svecia
Bearded Tit
  Panurus biarmicus
Penduline Tit
  Remiz pendulinus
River Warbler
  Locustella fluviatilis
Sedge Warbler
  Acrocephalus schoenobaenus
Great Reed Warbler
  Acrocephalus arundinaceus
Reed Warbler
  Acrocephalus scirpaceus
Savi's Warbler
  Locustella luscinioides

A home for innumerable European waterbirds, Austria's Lake Neusiedl is well known for its white storks (top) that nest on village chimneys throughout the region. Their elaborate greeting ceremonies are accompanied by a loud bill clapping. Below, the red-necked grebes build their floating nests among the shallows.

Ngorongoro Crater Conservation Area, Tanzania

Variously described as the greatest wildlife show on earth and the eighth wonder of the world, Ngorongoro is virtually beyond description. It has been filmed and photographed almost *ad nauseam* and yet still attracts thousands of wildlife specialists including the professionals, every year. To see Ngorongoro in their lifetime is one of the major aims of every naturalist.

The volcanic crater, or more correctly the caldera, lies in northern Tanzania to the east of Lake Victoria and the vast plains of the Serengeti. It is the sixth largest caldera in the world, but the largest one that is unbroken and unflooded. It is ten to twelve miles in diameter and the floor covers 102 square miles at an average altitude of 5600 feet. The rim of the caldera rises a further

Perhaps the greatest of all wildlife
paradises, the Ngorongoro Crater teems with
animals and their predators. A male lion
(below) drags a zebra towards cover where
it can eat in peace. Even the black
rhinoceros (left) is in danger while it is
young and remains in close contact with
its mother.

2000 feet, though within the 3200 miles of the Ngorongoro Conservation Area the land rises to 11,769 feet at the peak of Lolmalasin.

Looking down into the crater from the rim is an experience of a lifetime. Thousands upon thousands of large animals spread out across the plains which are broken only by a few streams, marshes and lakes. A census in 1964 produced 22,000 head of game on the crater floor, while on one occasion or another no less than fifty-five species of mammals have been recorded.

Though the grassland animals dominate, there is such a variety of habitat that the inquiring visitor can pick up a whole range of animals that are seldom seen by the tourist day-tripper. Lion, elephant, black rhinoceros, hippo and buffalo can all be found without difficulty along with the commoner gazelles, eland, zebra, wildebeest, cheetah, spotted hyena and large numbers of hunting dogs that are difficult to see elsewhere. Giraffe, waterbuck, hartebeest, warthog all fall to the photographer tourist but leopard, hyrax, giant forest hog and bush duiker are usually missed.

The variety of habitats has lead to a check-list of over 340 species of birds in the Conservation Area. Ostrich, kori bustard, blacksmith and crowned plovers and the numerous crowned crane are found throughout the area alongside the mammals. There are six vultures, twenty-seven birds of prey, all the usual herons and egrets, five species of stork, and very often large flocks of flamingoes on the lake. The more dedicated ornithologists will find the woodlands at Lerai and Laindi particularly rewarding for the smaller and less obvious species of birds.

Human occupation of the crater is not banned and the colourful, pastoral Masai are encountered as they herd their cattle across the plains. Like the wild animals that they live alongside, they do not object to photography, though they do expect payment.

Visiting: Lies to the west of Arusha along the main road past Lake Manyara. Access to the crater itself is only allowed to four-wheel drive vehicles. Visitors without such facility can join parties at Crater Lodge and the other lodges and at the beginning of the nearby Lerai Descent. Accommodation at the Lodge is luxurious but visitors can stay at the more modest Forest Lodge and at Kimba Tented Camp.

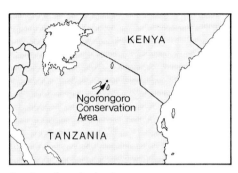

Species of particular interest

Lion
  Panthera leo
Cheetah
  Acinonyx jubatus
Elephant
  Loxodonta africana
Black Rhinoceros
  Diceros bicornis
Buffalo
  Synceros caffer
Hippopotamus
  Hippopotamus amphibius
Giraffe
  Giraffa camelopardalis
Wild Dog
  Lycaon pictus
Spotted Hyena
  Crocuta crocuta
Wildebeeste
  Connochaetes taurinus
Zebra
  Equus burchelli
Thomson's Gazelle
  Gazella thomsoni
Grant's Gazelle
  Gazella granti
Eland
  Taurotragus oryx
Coke's Hartebeest
  Alcelaphus buselaphus
Waterbuck
  Kobus ellipsiprymnus
Impala
  Aepyceros melampus
Bushbuck
  Tragelaphus scriptus
White Pelican
  Pelecanus onocrotalus
Pink-backed Pelican
  Pelecanus rufescens
Saddle-bill Stork
  Ephippiorhynchus senegalensis
Secretarybird
  Sagittarius serpentarius
Black-shouldered Kite
  Elanus caeruleus
Verreaux's Eagle
  Aquila verreauxii
Martial Eagle

  Polemaetus bellicosus
Lammergeier
  Gypaetus barbatus
Crowned Crane
  Stephanoaetus coronatus
Helmeted Guineafowl
  Numida mitrata
Variable Sunbird
  Cinnyris venustus
Grey Hornbill
  Tockus nasutus

During the rutting season male impalas stake out territories in the hope of attracting a herd of females. Boundary fights are commonplace.

Okavango Swamps, Botswana

The Okavango River rises in the highlands
of Angola and flows south-eastwards into
Botswana where it divides into a number
of channels and floods over the Okavango
Swamps. Here is an inland delta, a vast
network of inter-connected channels
and waterways that is unique in the dry
interior of southern Africa. Into this delta
pour 6000 million gallons of water every
day. At times the water is more than the
delta can cope with and it overflows
southwards towards the evaporation pan
of the Makarikari Depression, or north-
eastwards towards the great Zambesi
via the Selinda and Linyati channels.
Water in these channels sometimes flows
one way and sometimes the other. It is a
peculiar arrangement, but it does mean
that the Okavango Swamps remain fresh
with all of the life forms that hot,
tropical swamps support.

Vast beds of reeds and papyrus cover
the floods, broken only by the islands of
solid land that stand up between the arms
of the channels. One of these has been
declared the Moremi Wildlife Reserve by
the tribal owners of Ngamiland and
covers 700 square miles of prime wildlife
habitat. The area includes swamps as well

The Okavango Swamps teem with life.
Lechwe (right) bound through the shallow
waters. Characteristic of the drier parts of
Okavango, steenbok (left) prefer the
concealing grasslands of this remote and
inaccessible area.

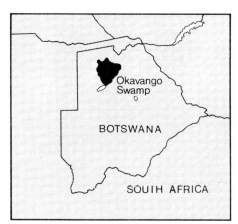

as grassland and woodland. It supports over fifty species of large mammals and is as rich in birdlife as any comparable area in Africa. Even Kenya's Lake Naivasha is only equal, and certainly not superior, to the swamps of Okavango.

Over most of the swamps water lilies grow in profusion forming a home for the African jacana and lesser jacana, both often jointly referred to as lily-trotters. The black crake creeps among the vegetation while a whole host of African herons and egrets vie for attention. Wattled cranes engage in their elaborate courtship, while black herons create umbrellas from their wings under which to hunt their prey. Saddlebill and yellow-billed storks, sacred and hadada ibises and no less than eleven other species of herons can all be found. Each has its own distinctive method of feeding, a method that separates it from its numerous competitors in some way. Seeing a line of herons and egrets along an Okavango channel leads one to wonder how these birds are separated, but there are fish in plenty for all. Catfish and tilapia are so numerous and so large that it is not surprising that there are so many fish eagles, but a range of other raptors can be seen as well. Kingfishers too are well represented, with five species all in action.

Of the large mammals the red lechwe are outstanding as they leap gracefully across the splashy lagoons. They are difficult to approach and photograph but few of the other animals are so timid. Hippopotami are numerous and are of great importance to the ecology of the Swamps in keeping the channels open, a function that dredgers find progressively more difficult. Buffalo and crocodiles too find the swamps to their liking but there are a host of other large mammals that are more generally thought of as being savannah dwellers and which enjoy refuge on the islands. Elephant, lion, leopard,

cheetah, roan, sable, eland and wildebeest can all be found.

This is a paradise that is still virtually unknown to the outside world, but it is also a paradise in danger. There are schemes to harness the water of Okavango to supply diamond mines to the south and even to sell to South Africa. It is difficult for a poor nation like Botswana to resist the temptations that such economic developments offer.

Visiting: The Botswana tribe has appointed guards and established the Moremi Reserve as a wildlife park capable of receiving visitors. But for most of the Swamps the visitor is on his own. He must find his way to the town of Maun and hire craft locally to explore this vast area.

Species of particular interest

Lion
 Panthera leo
Leopard
 Panthera pardus
Cheetah
 Acinonyx jubatus
Elephant
 Loxodonta africana
Hippopotamus
 Hippopotamus amphibius
Red Lechwe
 Kobus leche leche
Steenbuck
 Raphicerus campestris
Reedbuck
 Redunca redunca
Waterbuck
 Kobus ellipsiprymnus
Roan
 Hippotragus equinus
Sable
 Hippotragus niger
Tsessebe
 Damaliscus lunatus
Bushbuck
 Tragelaphus scriptus
Kudu
 Tragelaphus imberbis
Buffalo
 Syncerus caffer
Eland
 Taurotragus oryx
Zebra
 Equus burchelli
Side-striped Jackal
 Canis adjustus
Silver-backed Jackal
 Canis mesomelas
Bat-eared Fox
 Otocyon megalotis
Ostrich
 Struthio camellus
Darter
 Anhinga rufa
Goliath Heron
 Ardea goliath
Black Heron
 Melanophoyx aedesiaca
Skimmer
 Rynchops flavirostris
Pygmy Goose
 Nettapus auritus
Cape Vulture
 Gyps eoprotheres
African Fish Eagle
 Haliaetus vocifer
Bateleur
 Terathopius ecaudatus
Martial Eagle
 Polemaetus bellicosus

*Olive baboons and many other mammals
find refuge in the security of the swamps.*

Pribilof Islands, Alaska, USA

Isolated and lying in the middle of the Bering Sea are the Pribilof Islands. The name alone conjures up visions of an isolated, fog-covered group of rocks, echoing to the cries of seabirds—and so it is. Here in the cold, sub-Arctic waters life is plentiful and it is rich enough to support one of the greatest concentrations of mammals on earth.

The Pribilofs lie nearly 300 miles off the coast of mainland Alaska and a similar distance north of the nearest of the Aleutian Islands. Though there are five islands the two major ones are about forty miles apart. St George, in the south, has the largest seabird colonies but is more difficult of access and accommodation. St Paul, on the other hand, is a little larger, boasts a larger human population, much the same seabirds and the largest rookery of Alaskan fur seals. The east coast of St Paul becomes alive every May with two to three million fur seals. The bulls, the 'beachmasters', haul up and surround themselves with as large a harem as possible. The noise, smell and extraordinarily restless sight is either thrilling or sickening depending on the visitor's stomach. Within a few days of giving birth to their pups the cow fur seals are mated and so the procreation process is continued the following year.

Though these huge colonies of fur seals are an impressive sight there are other attractions on the Pribilofs. The cliffs teem with a variety of alcids (called auks in Europe) that is unequalled anywhere in the world. The common and thick-billed murres (guillemots) are both abundant but there are good numbers of horned and tufted puffins and truly vast numbers of parakeet and crested auklets. Kittiwakes are common, but there are two species here and bird-watchers are pleased to see the red-legged as well as the more widespread black-legged. Other gulls are seldom plentiful but American birders get great thrills from seeing many Eurasian (Palearctic) birds on their own continent.

Arctic species such as harlequin duck, Steller's and king eiders, Lapland longspurs and snow buntings are commonly seen and no less than 170 species have been noted in total.

Visiting: Not very easy to visit. The flight from Anchorage to St Paul is expensive. There is a single hotel which is clean, warm and comfortable, but the climate is cold, wet and foggy.

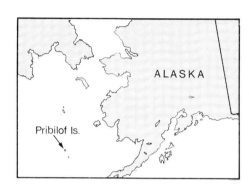

ALASKA

Pribilof Is.

Species of particular interest

Alaskan Fur Seal
  Callorhinus ursinus
King Eider
  Somateria spectabilis
Harlequin Duck
  Histrionicus histrionicus
Red-faced Cormorant
  Phalacrocorax urile
Common Murre
  Uria aalge
Thick-billed Murre
  Uria lomvia
Parakeet Auklet
  Cyclorrhynchus psittacula
Crested Auklet
  Aethia cristatella
Horned Puffin
  Fratercula corniculata
Tufted Puffin
  Lunda cirrhata
Pigeon Guillemot
  Cepphus columba
Kittiwake
  Rissa tridactyla
Glaucous Gull
  Larus hyperboreus
Grey-crowned Rose Finch
  Leucosticte tephrocotis

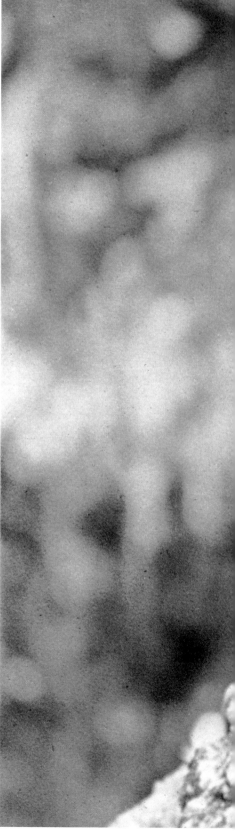

The remote Pribilof Islands are a major resort for
a variety of auks found only in the northern Pacific.
These tufted puffins have the boldly coloured bill
typical of their tribe.

Cape Romain National Wildlife Refuge, South Carolina, USA

The North American continent enjoys four quite distinct bird migration flyways–the west coast, the central, the Mississippi and the east coast. All have an individual character and species that are unique to them, but because of the uniformity of their origins, in the great belts of taiga and tundra, each has much in common with the others. More than anything else they share geese, thousand upon thousand of geese. From the Sacramento Valley in the west through Sand Lake in the central, to Cape Romain in the east, geese are the common denominator.

Covering 60,000 acres Cape Romain National Wildlife Refuge is noted for the concentrations of Canada geese, with smaller numbers of blue and snow geese, that appear here every winter. With them come quite uncountable numbers of duck and shorebirds to make the area one of the greatest areas for migratory birds in North America.

Cape Romain consists of salt marshes and inter-tidal areas of sand and mud, together with famous Bull's Island. This low-lying island covers nearly 5000 acres and is the only part of the Refuge to boast woodland. It is also the best area for wildfowl and is almost entirely given over to the production of food to attract and retain these species. Black duck, pintail, green and blue-winged teal, shoveler, mallard, gadwall, ruddy duck, scaup, ringed-necked duck and bufflehead are all found by the hundred, if not by the thousand. Bull's Island manages to support a considerable number of these species throughout the year and has a total duck list of twenty-two species.

Beyond Bull's Island, among the maze of offshore islands and shoals, there are brown pelicans, the large orange-billed royal terns, oystercatchers, willets, least terns and skimmers by the hundred. Oystercatchers are particularly attracted to Cape Romain–the only place on the eastern seaboard where they winter in any number. The lean season brings hosts of shorebirds to the Refuge including long-billed curlew and marbled godwit. Least and western sandpipers, dowitchers, plovers, turnstones, all feed on the open mud flats. Forster's and Caspian terns stay during the winter, the climate is so mild.

Though the larger aquatic birds are so obvious and dominating there are plenty of smaller fry to attract more dedicated birders. Red-cockaded woodpecker, white-eyed and red-eyed vireos, painted bunting, clapper rail, Carolina wren,

brown-headed nuthatch, all can be found within the area along with a host of others, including some that are not small–like the wild turkey.

As in all good wetland refuges waterfowl are the main attraction, but there are also some quite interesting terrestrial animals including alligator, raccoon, white-tailed deer and fox squirrel.

Visiting: Take the Sewee road off US Route 17 to Moore's landing. Here arrangements can be made to explore the refuge and visit Bull's Island.

Species of particular interest

White-tailed Deer
  Odocoileus virginianus
Raccoon
  Procyon lotor
Fox Squirrel
  Sciurus niger
Alligator
  Alligator mississipiensis
Canada Goose
  Branta canadensis
Pintail
  Anas acuta
Gadwall
  Anas strepera
Bufflehead
  Bucephala albeola
Brown Pelican
  Pelecanus occidentalis
Black Skimmer
  Rynchops nigra
Willet
  Catoptrophorus semipalmatus
Marbled Godwit
  Limosa fedoa
Long-billed Curlew
  Numenius americana
Least Sandpiper
  Calidris minutilla
Western Sandpiper
  Calidris mauri
Oystercatcher
  Haematopus ostralegus
Royal Tern
  Sterna maximus
Forster's Tern
  Sterna forsteri
Caspian Tern
  Hydroprogne tschegrava
Elegant Tern
  Thalesseus elegans
Clapper Rail
  Rallus longirostris
Red-cockaded Woodpecker
  Dendrocopus borealis
Painted Bunting
  Passerina ciris
Chuck-will's Widow
  Caprimulgus carolinensis
Carolina Wren
  Thryothorus ludovicianus

Ruwenzori National Park, Uganda

Formerly known as the Queen Elizabeth National Park, Ruwenzori takes its name from the mountains of the same name— the mythical 'Mountains of the Moon'. The Park covers over 700 square miles in the extreme west of Uganda against the border with the Republic of Zaire, formerly known as the Congo. It is this geographical position far to the west of what is generally regarded as 'East Africa' plus a wide variety of habitats that gives Ruwenzori its unique character. In particular the centre of the Park is occupied by tropical forest, the Maramagambo, more in keeping with the forests of the Congo than with the rest of East Africa with its open savannahs and acacias.

Though gorillas do not penetrate the Park there are a host of other primates present in Maramagambo. Chimpanzees are an obvious attraction and family groups can usually be observed from the road that cuts through the forest

Ruwenzori is an East African paradise boasting a huge range of species. The hippopotamus (left) is particularly numerous here and, while great photographs are never easy to obtain, these often shy animals are usually seen to advantage. Buffalo (below) occur in large herds, while the warthog (bottom) can often be seen in drier and more open areas.

connecting the plains to the north and
south. While red colobus monkeys are
extremely rare there is always a good
chance of seeing the black and white
colobus, the blue monkey and the
black-faced vervet monkey. The forest
zone also supports a wide variety of birds
including many sunbirds and woodpeckers.
The lakes to the north-east of the forest
are a renowned bird locality boasting
superb kingfishers as well as storks and
ibises. There is always a fair chance of the
boat-billed stork in Ruwenzori. But the
Nyamagesani River is perhaps the best
place of all for water birds. Storks and
ibises here are numerous and of a wide
variety of species. There are duck and
geese and the area is also rich in
scavenging vultures and the larger eagles
and hawk-eagles. European waders are
often numerous.

Among the larger plains game there are
several notable absentees from the more
usual safari list, but elephant and buffalo
are plentiful and there are usually large
numbers of kob and defassa waterbuck.
Lion and leopard are quite common but
are never as obvious as in the parks to
the east like the Serengeti. There are over
10,000 hippopotami in the Park, a
concentration that is quite unique and
which leads to the most incredible jumble
of bodies in the shallows.

Because of its variety of habitats
Ruwenzori is a quite excellent place for
bird-watching and an experienced
observer usually has no difficulty in seeing
a hundred species in a day. Nevertheless
even the experienced find some difficulty
in separating five species of honey-guide,
four cisticolas and some dozen species
of cuckoo.

Visiting: The Park lies some 300 miles
from Entebbe and can be reached by road.
The Safari Lodge at Mweya overlooks
Lake Edward and provides first-class
accommodation.

Species of particular interest

Lion
  Panthera leo
Leopard
  Panthera pardus
Chimpanzee
  Pan troglodytes
Red Colobus Monkey
  Colobus pennanti
Black and White Colobus Monkey
  Colobus polykomos
Black-faced Vervet Monkey
  Cercopithecus aethiops
Blue Monkey
  Cercopithecus mitis
Hippopotamus
  Hippopotamus amphibius
Elephant
  Loxodonta africana
Warthog
  Phacochoerus aethiopicus
Topi
  Damaliscus korrigum
Buffalo
  Syncerus caffer
Waterbuck
  Kobus ellipsiprymnus
Bushbuck
  Tragelaphus scriptus
Darter
  Anhinga rufa
Goliath Heron
  Ardea goliath
Black Heron
  Melanophoyx ardesiaca
Hammerkop
  Scopus umbretta
Shoe-billed Stork
  Balaeniceps rex
Woolly-necked Stork
  Dissoura episcopus
White-faced Tree Duck
  Dendrocygna viduata
Pygmy Goose
  Nettapus auritus
Palm-nut Vulture
  Gypohierax angolensis
Egyptian Goose
  Alopochen aegyptiacus

While hippos enjoy the restful cool of the water they provide an ideal perch for the curious hammerkops.

Samburu-Isiolo Game Reserves, Kenya

Samburu lies **180** miles north of Nairobi in
the Northern Frontier Province and in
the lands of the tribe of that name.
It's an off-the-beaten-track place not on
the tourist circuit and all the better for it.
It's a dry area but an intensely beautiful
one—a landscape that owes little to
modern civilization, where nomadic
people still roam as they have done since
before white men penetrated the
continent. Yet the Reserve is connected
by good motor roads to Nairobi and
there is a lodge offering luxury standard
accommodation.

Lying north and being so dry, Samburu
has different attractions to the southern
parks. Grevy's zebra are surprisingly more
appealing than the common Burchell's
and are found here in good numbers.
Reticulated giraffe replace the more
commonly encountered Masai giraffe, and
beisa oryx are widespread. This does not
mean that Samburu lacks the great stars
of an East African safari—lion, leopard,
elephant, black rhinoceros, hippopotamus
are all found here.

Samburu lies north of theUaso Nyiro
River and includes a ten-mile long stretch
of river bank. Across the river to the
south lies the Isiolo Game Reserve which
has similar animals and habitats. The two
Parks are connected via a single bridge
near the Samburu Lodge and visitors can
explore both from this base.

In this environment it is the river itself
that proves one of the major attractions.

The dry, arid savannahs of Samburu
Game Reserve hold large numbers of beisa
oryx (below) and gerenuk (left). But in the
heat of the day all the animals seek the
shade of the acacia trees.

Vast herds of game animals come to drink, along with huge flocks of birds from the surrounding arid regions. Guinea-fowl are particularly attractive and numerous though perhaps nothing could outshine the spectacle of thousand upon thousand of sandgrouse as they drink morning and evening at Buffalo Springs in Isiolo. While most reserves offer a true wealth of birds, few can offer a bird sight of such dimensions.

Bird-watching is particularly good along the river, though the open plains hold the Somali ostrich, three species of courser and three of sandgrouse, several bustards and many plovers. Among other species that the visitor will be pleased to see are cheetah, leopard, gerenuk, lesser kudu, the incredibly tame ground squirrel and two species of hyena.

Visiting: Easily reached by road from Nairobi. Most visitors take the route which includes Naivasha and Nakuru. Accommodation at the Samburu Lodge is luxurious and there are bandas and a camp site available.

Species of particular interest

Lion
  Panthera leo
Leopard
  Panthera pardus
Cheetah
  Acinonyx jubatus
Grevy's Zebra
  Equus grevyi
Reticulated Giraffe
  Giraffe camelopardalis reticulata
Beisa Oryx
  Oryx beisa
Lesser Kudu
  Tragelaphus imberbis
Waterbuck
  Kobus ellipsiprymnus
Gerenuk
  Litocranius walleri
Striped Hyena
  Hyena hyena
Elephant
  Loxodonta africana
Black Rhinoceros
  Diceros bicornis
Hippopotamus
  Hippopotamus amphibius
Somali Ostrich
  Struthio molybdophanes
Secretarybird
  Sagittarius serpentarius
Pygmy Falcon
  Poliohierax semitorquatus
Black-shouldered Kite

Elanus caeruleus
Verreaux's Eagle
  Aquila verreauxi
Martial Eagle
  Polemaetus bellicosus
Vulturine Guineafowl
  Acryllium vulturinum

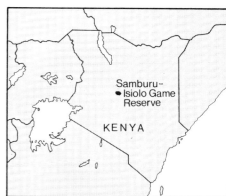

Grevy's zebra (below), to be distinguished from the more broadly striped Burchell's, is common at Samburu, though it seldom forms large herds like that species. Even the long neck of the gerenuk (left) is not always sufficient to reach the more tasty shoots of the acacias. It is well adapted to partial two-footed existence.

Serengeti, Tanzania

There is no wildlife sight on earth to compare with the plains of the Serengeti. At all seasons it is second to none, but in February and March, when the wildebeest are on migration there are animals as far as the eye can see. The Serengeti National Park covers a huge area: 5700 square miles or about the same size as Connecticut or Northern Ireland. Yet for all this size the park is not a complete ecological unit; the game herds move outside its boundaries from time to time during their annual wanderings. Sometimes parts of the park seem empty of animals but from November through to May there are 350,000 wildebeest, 180,000 zebra, over half a million Thomson's and Grant's gazelle along with hosts of the other antelopes making a total of over 1,500,000 large mammals.

The Seronera area has been noted as a haunt of lions ever since an American, Mr L. Simpson, reached there in his model 'A' Ford in 1920. By 1929 the visitors and the slaughter had increased to such an extent that a 900 square mile sanctuary was established. Now lions roam freely along with leopards that are as numerous here as anywhere in Africa. Their daytime roosts in trees along the Seronera River are well known. Other carnivora include cheetah, all three species of jackal, striped hyena and hunting dog.

Elephant, black rhinoceros, hippopotamus and giraffe are all found commonly. Less frequently encountered are fringe-eared oryx, two species of waterbuck, steinbok, three duikers and so on.

Though altitude ranges only between 3000–6000 feet, there is considerable variation of habitat from the dominant long and short grass plains in the south and east to the clay plains of the mountains in the western corridor. There are rivers, lakes, hills and rocky outcrops called 'kopje', the latter providing a home not only for hyrax but also for Kirk's dikdik and klipspringer. Perhaps not surprisingly over five hundred different species of birds have been identified, and even a day confined to the area around Seronera should produce well over a hundred species for anyone with a familiarity with birds.

Throughout the area there are rollers and bee-eaters in colourful profusion. Ostrich and bustards roam the plains while the marshes and lakes provide a home for a host of waterbirds including darter, pelicans, goliath heron, green-backed heron, hammerkop, eight species of stork, all six vultures, the delightful

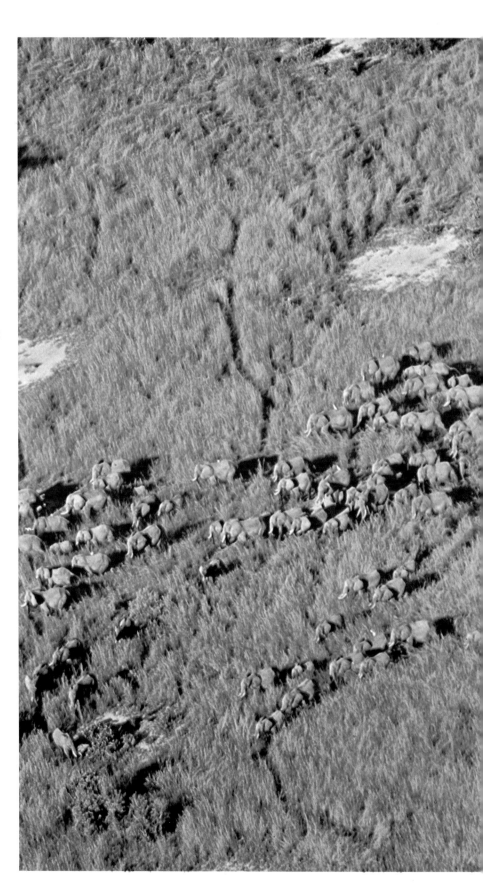

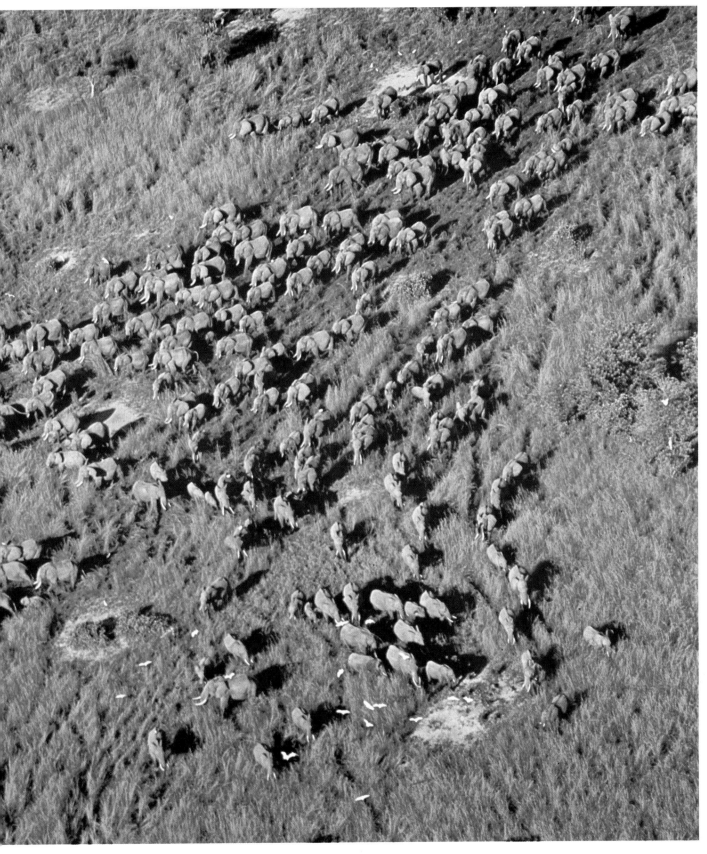

Vast herds of animals are everyday features of the Serengeti landscape. Seen from the air a herd of elephants swarms across the bush.

Drinking is an essential but dangerous part of the lives of all animals. Burchell's zebra (left) are ever watchful at a Serengeti waterhole. Leopards (right) are great climbers. They sleep out the day in trees, and even carry their prey aloft where it is safe from scavengers.

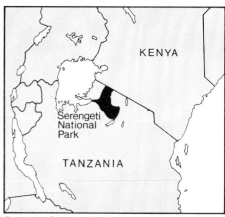

Species of particular interest

Cheetah
  Acinonyx jubatus
Lion
  Panthera leo
Leopard
  Panthera pardus
Wildebeeste
  Connochaetes taurinus
Burchell's Zebra
  Equus burchelli
Buffalo
  Syncerus caffer
Black Rhinoceros
  Diceros bicornis
Elephant
  Loxodonta africana
Topi
  Damaliscus korrigum
Eland
  Taurotragus oryx
Wild Dog
  Lycaon pictus
Striped Hyena
  Hyena hyena
Grant's Gazelle
  Gazella granti
Thomson's Gazelle
  Gazella thomsoni
Hartebeest
  Alcelaphus buselaphus
Hippopotamus
  Hippopotamus amphibius
Ostrich
  Struthio camellus
Darter
  Anhinga rufa
Hammerkop
  Scopus umbretta
Sacred Ibis
  Threskiornis aethiopicus
Verreaux's Eagle
  Aquila verreauxi
Wahlberg's Eagle
  Aquila wahlbergi
Bateleur
  Terathopius ecaudatus

pygmy falcon, ten species of eagles and bataleur resplendent in black and red plumage and often very approachable. There are francolins and spurfowl, bustards and sandgrouse, resident and migrant (European) waders which compete for attention with the local skimmers swishing up and down over the water in their own peculiar method of fishing. Seven species of kingfisher, six of hornbill, sixteen species of shrike and more weavers than you can identify in a day are other highlights in what can be a dawn till dusk orgy of bird-watching.

There are other animals like bats and snakes and a wealth of flowers so that all

tastes are catered for in this paradise that is called the Serengeti.

Visiting: Arrive by road from the south-east via Arusha and Ngorongoro. Seronera Lodge is the centre for visitors, who obviously cannot hope to see much more than a small proportion of the area. There are three camp sites by the Lodge as well as an airstrip in regular contact with Nairobi.

Seychelles, Indian Ocean

Ninety-five idyllic, desert islands strung
out across the Indian Ocean and washed
by the coral seas make up the Seychelles
group. On a map they appear as offshore
islands of the continent of Africa or
outliers of the great island of Madagascar.
In fact they are over 650 miles from each
and different enough to have a climate
and fauna of their own. Their names—
Mahé, Felicité, Aride, Curieuse—add an
air of romantic mystery to the desert
island theme and it is easy to imagine
some cut-throat band of pirates burying
treasure and hiding maps. Today the
islands are idyllic though 50,000 people
live on Mahé the largest of the group.

Of all the islands Cousin is one of the
smallest, only sixty acres in extent, and
richest. This tiny speck was purchased in
the late 1960s by the International
Council for Bird Preservation following
public appeals and the support of
thousands of naturalists. It lies off the
south-west coast of Praslin Island and is
the least spoilt of all the Seychelles.
A marsh, parts of the original jungle and,
despite commercial exploitation for
coconuts, much of the original fauna and
flora remain. As elsewhere in the group
Seychelles' endemic birds are a considerable attention to naturalists, though less
dedicated visitors might find the further
ranging seabirds a more dramatic sight.
Cousin seems dominated by the vast
flocks of greater and lesser noddy terns
and by the delicate white fairy terns with
their false black eye. These exquisite little
birds lay their egg actually on the branches
of trees with no attempt at nest building
and are delightful to watch as they flit to
and from their chosen tree. Some 15,000
pairs of each of these species breed along
with lesser numbers of brown-winged
terns. Ornithological pride of place must
go, however, to the Seychelles brush
warbler which is endemic to Cousin, and
the Seychelles magpie robin found only
on this and a few other small rocky islets.
Seven other birds are endemic to the
Seychelles and most have earned a place in
the Red Book of endangered species.

Giant tortoises (of a different genus to
the Galapagos tortoises) can be found on
Cousin as can green and hawksbill
turtles, which here find one of their few
safe refuges. Other islands in the group
have their own attractions. Comoro, for
example, boasts its own Red Book species
including a thrush and the blue pigeon.
Praslin has the black parrot and lesser
vasa parrot together with the extra-
ordinary coco-de-mer—a giant double
coconut with strange medicinal properties.

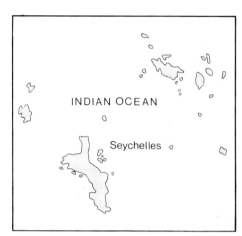

INDIAN OCEAN

Seychelles

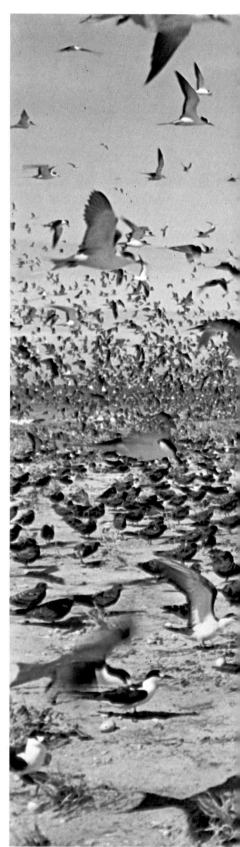

Mahé has the Seychelles Scop's owl and
La Digue the Seychelles black paradise
flycatcher which survives despite being
declared extinct in 1958.

Perhaps the greatest of all the islands of
the Seychelles is a tiny coral island in
the north—Bird Island. It holds an
estimated two million sooty and noddy
terns and unknown numbers of wedge-
tailed shearwaters that come awailing
in every evening during the season.

Bird spectacular, interesting
evolutionary site, superb under-water
and a great spot for those adding birds to
a life-list, that's the Seychelles.

Visiting: Regular air contact between
several parts of the world. Boats run
between the islands and visits can be
arranged to Cousin if prior application
is made to ICBP, c/o British Museum
(Natural History), Tring, Herts, UK.
Landing on the other islands is a matter
of arranging transport locally.

Species of particular interest

Giant Tortoise
  Testudo gigantea
Hawksbill Turtle
  Eretmochelys imbricata
Green Turtle
  Chelonia mydas
Lesser Noddy
  Anous tenuirostris
Greater Noddy
  Anous stolidus
Sooty Tern
  Sterna fuscata
Fairy Tern
  Gygis alba
Wedge-tailed Shearwater
  Puffinus pacificus
Seychelles Owl
  Otus insularis
Seychelles Brush Warbler
  Nesillas seychellensis
Seychelles Magpie Robin
  Copsychas seychellarum
Seychelles Sunbird
  Nectarinia dussumieri
Seychelles Fody
  Foudia seychellarum
Lesser Vasa Parrot
  Coracopsis nigra
Seychelles Black Paradise Flycatcher
  Tersiphone atrocaudata

Several isolated islands boast species that are not found elsewhere. The giant tortoise of Aldabra, a dependency of the Seychelles, has evolved in the security of a predator-free landscape.

Shetland, Great Britain

While it seems unlikely that one of the world's great nature paradises should be in Britain, that country's seabird colonies are among the very best, and visited by hordes of naturalists every year. Over a hundred islands are strung out just six degrees below the Arctic Circle in Shetland. Most are rugged and uninhabited while others support small communities of crofters and fishermen. There is something about a seabird island without people, but visitors could be forgiven if they decided to enjoy the comforts provided by the islands' hotels and farmhouses.

In the far north lies Unst, just large enough for a car to be useful. Here is the huge hill of Hermaness, a place of pilgrimage to many. Great skuas, called bonxies locally, have their stronghold here among the rolling peat bogs. Loose colonies are scattered over the moors with Arctic skuas between them. In the north and west the Atlantic rollers have carved steep cliffs that provide a home for a large gannetry, and the intrepid may climb down the cliffs to get among these magnificent birds. There are razorbills, guillemots, fulmars and kittiwakes in profusion and in some places puffins have honeycombed the cliff tops with their burrows. With thousands of birds wheeling below this is one of the finest natural phenomena in the world. There are other birds here too. Whimbrel and golden plover breed on the moorlands and the dashing merlin streaks by to grab some unsuspecting meadow pipit or twite.

There are large seal rookeries on several of the offshore rocks and grey, as well as common seals, frequently haul up on the more remote beaches and rocks.

To the south lies Fetlar, the green island of Shetland. It too boasts excellent seabird colonies but is renowned as the home of Britain's rarest bird—a single pair of snowy owls has bred since 1967. The birds are guarded day and night during the breeding season but visitors are guided to a special hide from which excellent views of the birds can be obtained. Fetlar is also famed for its red-necked phalaropes which are similarly protected.

The island of Noss is inhabited only by a single shepherd during the summer but visitors can see the magnificent six hundred foot high cliffs of the Noup of Noss from the boats that regularly circle the island from the nearby mainland at Lerwick. In fact tours by boats are among the best ways of seeing the full drama of the Shetland seabird cliffs.

To the south of Shetland proper lies Fair Isle, a lonely stack separated from all other land by overy twenty miles of open sea. It has breeding seabirds but is better known as the site of the Fair Isle Bird Observatory, famed for the rare birds that it attracts each autumn. At no other place in the world are the birds that one will see so unpredictable.

Visiting: In regular contact with the Scottish mainland by sea and air. The airport at Sumburgh is the hub of the inter-island air network. There are airfields at Unst and Fetlar and on several other islands. Accommodation must be booked in advance, except in Lerwick, and well in advance for Fair Isle in autumn.

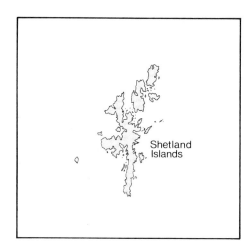

Species of particular interest

Common Seal
  Phoca vitulina
Grey Seal
  Halichoerus grypus
Great Skua
  Stercorarius skua
Arctic Skua
  Stercorarius parasiticus
Whimbrel
  Numenius phaeopus
Red-necked Phalarope
  Phalaropus lobatus
Gannet
  Sula bassana
Red-throated Diver
  Gavia stellata
Puffin
  Fratercula arctica
Razorbill
  Alca torda
Black Guillemot
  Cepphus grylle
Snowy Owl
  Nyctea scandiaca

The puffin is fast declining throughout its range, but is still a common sight in Shetland. Like penguins they stand about in groups apparently contemplating the sea.

Simien, Ethiopia

The Simien mountains lie in northern Ethiopia bounded by the rivers Takazza and Angareb which join the Nile below Khartoum via the River Angareb. With an average altitude of 10,000 to 13,000 feet they form a vast table-land disected by huge gorges that are thought by many to rival the Grand Canyon of Colorado in beauty. In several places sheer drops of 6000 feet are an awesome if impressive sight. In the north-east the plateau slopes down to 5000 feet with outlying pinnacles and flat topped 'ambas' beyond. It is a land that nature has created, a spectacular land of incredible harshness and beauty.

Among the rocky gorges there is a unique fauna and flora that includes several species found nowhere else in the world. St John's Wort grows to tree size while the giant lobelia has spikes that grow grotesquely to over twenty feet. On the faces of the gorges themselves Walia Ibex fleet sure-footedly over the rocks. This is a distinct species from the more northerly Nubia ibex which is found in the hills along the Sudan Red Sea and in Eritrea. The Walia lives between 8000 and 11,000 feet and in two major areas—between San Gabor and Amba Ras and to the north-east at Silki, Abba Yared and Walia Kand. Population estimates vary between 150 and 400 and there is no doubt that only really effective conservation measures will save this species from extinction.

In contrast the Gelada baboon, which shares the Walia's habitat, is both widespread and present in fair numbers, though its main stronghold is Simien. It forms small social groups with a definite hierarchical structure dominated by the senior male. They feed on roots and take to the cliffs at the approach of danger and for sleeping. With few natural enemies apart from the declining leopard they are causing considerable damage to crops and may have to be controlled in future.

The Simien fox looks like a cross between the European fox and the North American coyote. Its numbers are declining rapidly in Simien though it is still quite numerous in the Arussi and Bale mountains to the south. It is found between 11,000 and 13,000 feet and, despite accusations by shepherds, lives almost exclusively on the large colonies of rats found at these altitudes.

Klipspringer, bushbuck, duiker and other mammals are found in Simien along with a wealth of interesting birds. The lammergeier is particularly numerous and obvious and other birds of prey find

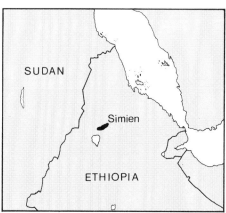

the living equally to their satisfaction. Augur buzzards are common and both fan-tailed and thick-billed ravens soar along the rocky gorges.

Visiting: Not easy to visit. Roads and accommodation are poor but an increasing number of wildlife safari companies are organizing tours in the region. Independent arrangements can be made in Addis Ababa. The best jumping-off point is Deborek, some three hours drive from Addis.

Species of particular interest

Walia Ibex
  Capra ibex walie
Simien Fox
  Canis simiensis
Gelada Baboon
  Papio gelada
Leopard
  Panthera pardus
Klipspringer
  Oreotragus oreotragus
Bushbuck
  Tragelaphus scriptus
Duiker
  Cephalophus spp.
Lammergeier
  Gypaetus barbatus
Tawny Eagle
  Aquila rapax
Thick-billed Raven
  Corvus crassirostris
Fan-tailed Raven
  Rhinocorax rhipidurus
Augur Buzzard
  Buteo rufofuscus

The high precipices of Simien offer security
to the unique gelada baboon (left), the males
of which boast a long, lion-like mane.
Along the sheer faces lammergeiers (right)
glide in numbers seldom found elsewhere.

Taman Negara National Park, Malaysia

Formerly known as King George V National Park, Taman Negara is situated in peninsular Malaysia near the border with Thailand. It is a vast area centred on the slopes of Gunong Tahan, at 7180 feet the highest mountain on the mainland. Dense, tropical forest covers the ridges and valleys making travelling very difficult and exploration a matter of organizing expeditions. But it is this very impossibility of transportation that has preserved the area from the agricultural development and monoculture that has ruined the rest of the country. Now there are opportunities for tourism and while this currently remains romantic and appealing, no doubt things will change in the near future.

At present the Park is roadless and all movement is by boat or foot. Among the huge trees and thick undergrowth there is a profusion of flowers, birds and mammals. For many species the area is of considerable importance on a world scale. Though pride of place must go to the Sumatran rhinoceros, visitors would be extremely fortunate to see that animal. It is the smallest of the world's rhinos and unlike the other Asiatic species boasts two horns. The world population is estimated at between 100 to 170 animals widely scattered over the area from Thailand through Malaysia to Borneo, and Sumatra where the highest concentration (about 60) is located. In Malaysia the remaining rhinos are so widely scattered that they must be doomed, except perhaps in the sanctuary of Taman Negara.

That other hyper-rare Asiatic wildlife star, the tiger, is also found in the forests. The Park authorities have organized tree-top hides to enable visitors to spend the night at Jenut Kumbang where there is a possibility of seeing the tiger as well as other rare and secretive animals like the tapir, Malayan gaur, otherwise known as the seladang, and black panther.

Even during the day there is a good chance of observing wild elephant in the Park and there are quite substantial herds of barking deer, sambar and wild pig. Many of these, as well as tapir, come to salt-licks and water holes that have been created so that visitors can see the wildlife of this unique area. The Javan bantung, though rare throughout its range, may still be found in the more open parts of the area. It is easily recognized by its white ankles.

Among the forests there are communities of native peoples living a primitive existence though they are not

adverse to selling the odd momento to the visitor—mainly bamboo pipes.

The bird and butterfly fauna is extremely rich and varied. Though the jungle is so dense, many birds can be found around the Park headquarters that are typical of the rest of the area. These include the lesser fishing eagle, large pied hornbill, blue-throated bee-eater, the brown-throated sunbird and the elusive broadbills. Their forest habitat is difficult to penetrate and the birds are self-effacing. Nevertheless, the green broadbill can be found by the patient observer! Further exploration should produce over a hundred species of which the most notable are: red-breasted falconet, Malay peacock and great argus pheasants, the serpent eagle and no less than fifteen species of bulbuls.

Visiting: The only way to get there is by road from Kuala Lumpur to the Park boundary and thence by boat up river for three hours. The Resthouse at Kuala Tahan is comfortable and has a bar, sanitation and filtered water. The Park can only be visited with the cooperation of the authorities at Kuala Tahan, Taman Negara National Park, Malaysia.

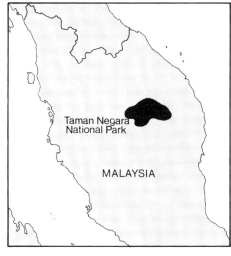

Taman Negara National Park

MALAYSIA

Taman Negara is not a one species park. The Java bantung (left) can be found among its forests, while bird life is prolific. The green broadbill (below) is surprisingly one of the less colourful members of its avifauna.

Species of particular interest

Sumatran Rhinoceros
  Didermocerus sumatrensis
Tapir
  Tapirus indicus
Tiger
  Panthera tigris
Black Panther
  Panthera pardus
Malayan Gaur
  Bos gaurus
Javan Banteng
  Bos banteng
Sambar
  Cervus unicolor
Barking Deer
  Memtiacus muntjak

Elephant
  Elephas maximus
Wild Pig
  Sus scrofa
Lesser Fishing Eagle
  Icthyophaga nana
Serpent Eagle
  Spilornis cheela
Fish Owl
  Bubo ketupa
Blue-throated Bee-eater
  Nyctyornis athertoni
Malay Peacock Pheasant
  Polyplectron malacense
Great Argus Pheasant
  Argusianus argus
Blue-crowned Hanging Parakeet
  Loriculus galgulus

Green Broadbill
  Calyptomena viridis
Fairy Bluebird
  Irena puella
Racquet-tailed Drongo
  Dicrurus paradiseus
Paradise Flycatcher
  Tersiphone paradisi
White-crowned Forktail
  Enicurus leschenaulti

Texel, Netherlands

The southern shore of the North Sea consists of the vast and treacherous inter-tidal shoals of the Waddenzee—the greatest haven for wading birds in Europe and possibly in the world. Over half a million birds find a living between the tides, though Dutch expertise is already casting its greedy eyes over the banks and shallows with a mind to enclose and drain this unique area. The job would be made easier by the chain of islands, the Friesian Islands, that run from the Dutch coast to south-western Denmark. All of the islands have been created by the sea and are a combination of sand and marsh; all are good for birds but none more so than Texel, the furthest west of all.

Texel lies only fifty miles north of Amsterdam and is connected to the mainland by a regular, cheap and frequent ferry service. It boasts nineteen distinct bird reserves and is one of the very few places in temperate western Europe that could class itself as a wildlife paradise. In the north the sea has created a huge dune system while to the south the landward side consists of a large area of saltings. Between the two, in a sort of saucer, lies the heart of Texel, the low-lying rich grass meadows beloved of waders.

All around Texel there are birds at all seasons, but in summer the reserves are alive with them, their ringing calls echoing over the flat marshes. Ruff and black-tailed godwit are the most obvious inhabitants of the fields though redshank are decidedly more numerous. Marsh and Montagu's harriers quarter the reed beds while black-necked grebes, bitterns and many duck scamper over the water into hiding.

The shore is a home to large numbers of terns and gulls with perhaps Sandwich terns being the most attractive and certainly the most densely packed. Kentish plovers trip along the beaches and avocets find suitable spots for their nests along the bare dune creeks. At the reserve of Muy, as well as at Geul, the most northerly spoonbills in Europe can be found nesting on the willows and elders that grow among the reeds.

In winter vast flocks of dunlin and knot, as well as geese and other wildfowl, find refuge on Texel, but during the peak migration seasons almost every wader of regular European occurrence can be found by those with the patience and skill to search. Turnstone, sanderling, little stint, curlew, sandpiper and a host of others are there for the identifying.

Visiting: Permits must be obtained for visiting most of the reserves but almost every hotel and guest-house on the island knows the routine and can advise. There is an excellent tourist information centre at VVV Texel, Holland.

Species of particular interest

Bittern
  Botaurus stellaris
Marsh Harrier
  Circus aeruginosus
Montagu's Harrier
  Circus pygargus
Kentish Plover
  Charadrius alexandrinus
Black-tailed Godwit
  Limosa limosa
Ruff
  Philomelos pugnax
Avocet
  Recurvirostra avosetta
Common Tern
  Sterna hirundo
Sandwich Tern
  Sterna sandvicensis
Herring Gull
  Larus argentatus
Common Gull
  Larus canus
Short-eared Owl
  Asio flammeus
Great Reed Warbler
  Acrocephalus arundinaceus
Icterine Warbler
  Hippolais icterina
Golden Oriole
  Oriolus oriolus

*One of the most outstanding bird reserves in
northern Europe, Texel Island holds large
numbers of breeding ruffs, the males of
which are gaudily plumed in spring.*

Tsavo National Park, Kenya

Tsavo, home of over 10,000 elephants, is the location for all of those pictures of rust-red jumbos. Tsavo is *the* place for elephant and in the sanctuary of its 8034 square miles they grow as big as anywhere except Marsabit, that lonely hill in north-eastern Kenya where cars must travel in pairs and where the old tuskers are the largest in the world. In Tsavo vast herds of elephants trample across the landscape, particularly to Mudanda Rock where hundreds of these magnificent beasts gather during the dry season to drink. Elephant also come to drink at the Aruba Dam which is twenty miles from Voi, and where their companions are often really large herds of buffalo.

Tsavo is a vast area. Within its boundaries lie desert, scrub, savannah, acacia groves, ridges and outcrops, hills and riverine forest—a complete cross-section of the East African scene. It is divided into the East and West Tsavo National Parks by the main Nairobi-Mombasa road and railway. So far only a small area of the Park has been opened up to travellers and the north-eastern and south-western sections can virtually be eliminated from everyone's plans. In fact the major attractions of Tsavo are centred in the West Park immediately south of the Nairobi-Mombasa road. Three lodges are grouped round the famous Mzima Springs where hippo-potamus live in numbers. The water springs from a lava ridge and is so pure that these great animals can be clearly seen in their element. A sunken

Tsavo National Park, the elephant place, is seen from the balcony of Kilaguni Lodge, an ever changing wildlife panorama. Coloured by the red earth, a solitary bull (left) drinks at a waterhole. Burchell's zebra (below) drink warily against a parched background.

179

observation hide enables the animals as well as the resident shoals of barbel to be viewed underwater. At nearby Kilaguni Lodge the wildlife photographer need do no more than put down his drink to obtain pictures of elephant as they come to drink at the adjacent pool.

A trip to Lugard Falls in the East Park is frequently rewarding, with large herds of all the usual game animals of East Africa and particularly for black rhinoceros— another Tsavo speciality. This area along the Galana River is a good place for lesser kudu, the striped antelope that merges so well with its background and which is among the most beautiful members of its family.

Masai giraffe, Coke's hartebeest, impala, Grant's gazelle, gerenuk, fringe-eared oryx and eland can all be seen without too much difficulty. And there is a wealth of birds here too. Eight different species of hornbills can be found, several of which are more likely to find you. They gather to be fed at the lodges along with the white-headed buffalo weavers. All six vultures find the pickings to their satisfaction while several species of eagle are equally at home. Pigeons and doves, bee-eaters and rollers, bustards and ostriches, hosts of babblers, flycatchers, shrikes, sunbirds, weavers— in fact almost the whole ornithological fauna of Kenya can be found in Tsavo.

Visiting: Whilst there are landing strips at Voi, Aruba, Kilaguni etc., most visitors will come by road from Nairobi or Mombasa. Leaving the road east or west from Tsavo village is ideal and leads directly into the best game areas. Lodges, bandas and tented camps offer accommodation.

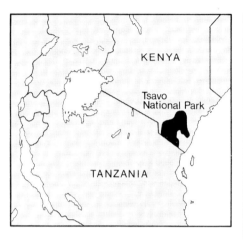

Species of particular interest

Lion
  Panthera leo
Leopard
  Panthera pardus
Cheetah
  Acinonyx jubatus
Elephant
  Loxodonta africana
Black Rhinoceros
  Diceros bicornis
Buffalo
  Synceros caffer
Lesser Kudu
  Tragelaphus imberbis
Masai Giraffe
  Giraffa camelopardalis
Coke's Hartebeest
  Alcelaphus buselaphus
Gerenuk
  Litocranius walleri
Fringe-eared Oryx
  Oryx beisa callotis
Masai Ostrich
  Struthio camellus
Somali Ostrich
  Struthio molybdophanes
Bat Hawk
  Machaerhamphus alcinus
Wahlberg's Eagle
  Aquila wahlbergi
Long-crested Hawk Eagle
  Lophoaetus occipitalis
Bateleur
  Terathopius ecaudatus
Pale Chanting Goshawk
  Melierax poliopterus
Peter's Finfoot
  Podica senegalensis
Kori Bustard
  Ardeotis kori
Chestnut-bellied Sandgrouse
  Pterocles exustus
White-headed Buffalo Weaver
  Dinemellia dinemelli

Coke's hartebeeste, sometimes called
kongoni, is found throughout Tsavo, though
it seldom strays far from water.

Wankie National Park, Rhodesia

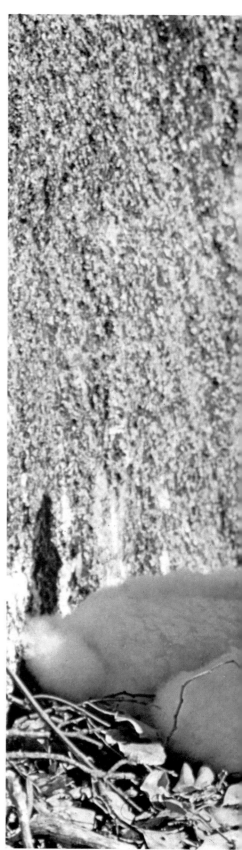

Covering 5540 square miles Wankie
National Park is the most important
wildlife area in Rhodesia and fit to rank
alongside other and better known parks
that lie to the north and south. The Park
is situated three hundred miles west of
Salisbury in that corner of the country
between Zambia and Botswana. To the
north lies the huge Zambesi River, with
Lake Kariba formed by the Kariba Dam
to the north-east. To the west, across the
border, is Okavango, one of the great
wetland areas of southern Africa.

Though on the fringe of the Kalahari
the area is not totally arid. Most of the
land is covered with scrub, but there are
quite large areas of forest. Water is not
plentiful and the Park authorities have
endeavoured to set up artificial waterholes that attract most of the game to
within camera distance. Similarly the
authorities have made successful efforts to
reintroduce species that were once
found there but which were exterminated by man. Black (about forty) and
white rhinoceros are the most obvious
examples.

All of the large African game animals
can be found at Wankie including
elephant. They number about 7000.
Small herds and family units gather at
waterholes and can frequently be
observed from hotel and lodge balconies.
Buffalo too spend much time at water
holes and often form large herds several
hundred strong. Lion, leopard, cheetah
and hyena can all be found along with
herds of antelopes many of which are not
found universally through East Africa.

182

Black-faced vervet monkeys (left) preen each other in small groups in Wankie National Park. Below, the magnificent Verreaux's or black eagle nests on ledges among inaccessible cliffs, but is frequently seen soaring overhead.

The sable forms decent sized herds and there are also roan in this area. Kudu, eland, waterbuck, impala, steenbuck and the arid-loving gemsbok can all be observed as can sassaby or tsessebe, a relative of the hartebeest. Giraffe, zebra and wildebeest are common, though not in the numbers found to the north on the plains of the Serengeti. Crocodiles haul out along the rivers and hippopotamus wallow in selected pools and waterways.

As would be expected birds are numerous, plentiful and obvious. Egrets, storks and herons are as common here as anywhere and there is an excellent collection of raptors. Martial eagle, Wahlberg's eagle, Verreaux's eagle, brown snake eagle, all can be found along with the usual vultures. Black-shouldered kites are not uncommon and auger buzzards often quite numerous. There are kingfishers over the pools and bee-eaters along the river banks, including some quite large colonies of carmine bee-eaters. Sunbirds are often common around the lodge grounds and some of the starlings, in their exquisite iridescent colours, come to feed on crumbs from the tea-table.

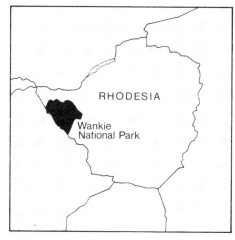

RHODESIA

Wankie
National Park

Visiting: The Park can be reached by regular services by rail and air from Salisbury or by road. There are three lodges offering simple but adequate accommodation on a self-catering basis and several camp sites. Only the northern part of the Park is open to tourists.

Species of particular interest

Lion
  Panthera leo
Leopard
  Panthera pardus
Cheetah
  Acinonyx jubatus
White Rhinoceros
  Ceratotherium simum
Black Rhinoceros
  Diceros bicornis
Buffalo
  Synceros caffer
Hippopotamus
  Hippopotamus amphibius
Side-striped Jackal
  Canis adustus
Hyena
  Hyena hyena
Roan
  Hippotragus equinus
Gemsbok
  Oryx gazella
Steenbuck
  Raphicerus campestris
Tsessebe
  Damaliscus lunatus
Crocodile
  Crocodylus niloticus
Black-shouldered Kite
  Elanus caeruleus
Carmine Bee-eater
  Merops nubicus

Restricted to south and west Africa, the roan
(left) is one of the most beautiful of antelopes.
In the background are Egyptian geese.
Carmine bee-eaters (below) bring a touch of
colour to the river banks. Though
somewhat localized, where they do occur
they are found in huge numbers.

Wood Buffalo National Park, Canada

Noted as the breeding site of the great, rare whooping crane, Wood Buffalo is a lot more beside. Covering 17,000 square miles it is the largest National Park in North America, a vast region of coniferous forests, marshes and bogs inter-connected by a maze of waterways that are confusing to the traveller and just as confusing from the air. The Park lies on the border between Alberta and the North-western Territories at a latitude where winter means heavy snow and hard frosts for months on end. When the thaw comes there is water everywhere, water that cannot run away. The low-lying marshes are then the home of the fifty odd whooping cranes that have spent the winter on the Gulf Coast of Texas at the Aransas Refuge, and whose progress northwards and southwards is watched with all the precision of a military operation.

Wood Buffalo Park lies between Lake Athabasca and the Great Slave Lake. It is a region that still retains its frontier character. The towns, which are few enough anyway, have a backwoods ring to their names—Fort Resolution, Salt River, Peace Point. Within the Park there is no accommodation, just what you bring yourself, and only one unmade road cuts through the area between Fort Smith and Hay River. It is ideal horse country, if you can find one.

Among the thousands of lakes the most obvious inhabitants are the birds. Most are migratory, wintering in the milder climes to the south, but the tough grouse manage to eke out a living through the lean season. In spring the meadows are alive with leks of sharp-tailed and ruffed grouse. It is then that the wildfowl arrive; thousands of duck come here to breed. With them come loons, sandhill cranes, hawks, ospreys and a whole wealth of sandpipers and peeps. Owls are among the first birds to breed and their calls echo through the empty forests of late winter.

The major factor in the establishment of Wood Buffalo as a National Park was the last remnants of the forest bison. In 1947 plains bison were introduced and the two sub-species now interbreed quite happily and number some 15,000 animals. Most of the other large mammals of the taiga, the zone where the coniferous forests meet the tundra, can be found within the Park. Moose find the mixture of water and woodland absolutely ideal and are numerous and widespread. Elk too do well and woodland caribou are found in quite sizeable herds. Both black and grizzly bears can be watched and there is a chance of seeing a wolf. Lynx, beaver, wolverine (glutton), porcupine all are found with diligent searching, or a well guided visit.

Wood Buffalo is a reminder of what the west was like before it was won. It is a great wilderness and a great refuge for wildlife. Though there is as yet little tourist or recreational development, the area has immense potential. Unfortunately there are plans afoot for a different form of development—plans that are unlikely to be put off by the simple device of a National Park.

Visiting: Trek, with the compulsory, but free, fire permit.

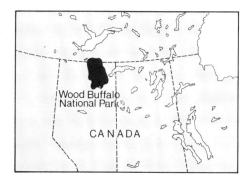

Wood Buffalo National Park

CANADA

One of the largest national parks in the world, Wood Buffalo is a vast wilderness area best known for its herds of bison (below). Bobcats (left) can be found among the rocky areas, but are no threat to the bison.

Species of particular interest

Forest Bison
   Bison bison
Plains Bison
   Bison bison
Moose
   Alces alces
Elk
   Cervus canadensis
Caribou
   Rangifer tarandus

Black Bear
   Ursus americanus
Grizzly Bear
   Ursus arctos
Wolf
   Canis lupus
Porcupine
   Erethizon dorsatum
Lynx
   Felis lynx
Beaver
   Castor fiber

Wolverine
   Gulo gulo
Whooping Crane
   Grus americana
Sandhill Crane
   Grus canadensis
Spruce Grouse
   Canachites canadensis
Ruby-throated Hummingbird
   Archilochus colubris

Yellowstone and Grand Teton National Parks, Wyoming, USA

In the north-western corner of the state of Wyoming, next to the Montana and Idaho borders, lies one of the most significant wilderness areas in the world. First penetrated by John Cotter in 1807-8, the outside world was sceptical of the wonders that he related and it was not until the official expedition of 1870 that Yellowstone was recognized as the natural wonder that it is. The story of how the expedition members sat round the camp fire discussing what should be done with the area is now a part of American folklore—the result was the world's first National Park created by the signature of President Ulysses S. Grant on 1 March 1872. Covering 3472 square miles at a height of between five to eleven thousand feet Yellowstone is the model for the whole world-wide National Park system, the very concept itself.

There is no doubt that Yellowstone and the Grand Teton a few miles to the south are great areas for mountain wildlife but their principle claim to fame must be the scenic wonders that are natural, though not living. Mountains carved into the most beautiful shapes by glaciers are an elegant background to the greatest volcanic show that most people are ever likely to see. Old Faithful, the most famous geyser in the world, lets rip as regularly as clockwork for the thousands of tourists that come to admire it every year. But there are hundreds of others, along with boiling mud, sulphur pools etc. They are the heart of the Park. Away from this area there are thousands of square miles that are seldom penetrated by the visitor—this is the home of the best of the Park's wildlife.

Though the black bears crowd round the cars at picnic sites, they are wild animals and can be dangerous. Away from the roads the visitor should know what he is doing for there are grizzlys and cougars here as well. The most obvious of the large mammals are the moose, elk, mule deer and bison, yet the star attractions are undoubtedly the bighorn sheep and pronghorn antelope, probably the fastest of all animals. These high level animals seem to evoke the true wildness of the area. Elk are large, attractive deer particularly numerous in Teton and in winter at Jackson Hole Wildlife Range, where winter feed is provided. Cats of three species, bobcat, lynx and cougar live in the Parks but are elusive— altogether some two hundred species of mammal have been recorded.

Of the birds the trumpeter swan is the undoubted star. Over half of the world population breeds in Yellowstone and Teton, and it is a very rare bird. Canada geese can be found on the lakes along with great blue herons. Golden and bald eagles, both increasingly scarce elsewhere, can still be found along with the decreasing sandhill crane.

Close relatives of the European chamois, pronghorn antelopes (below) grace the higher slopes of the Yellowstone Park, while from the high tops hoary marmots (left) look down to the valleys below.

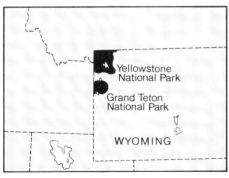

Species of particular interest

Black Bear
 Ursus americanus
Grizzly Bear
 Ursus arctos
Moose
 Alces alces
Bison
 Bison bison
Elk
 Cervus canadensis
Mule Deer
 Odocoileus hemionus
Pronghorn
 Antilocapra americana
Bighorn
 Ovis canadensis
Puma
 Felis concolor
Bobcat
 Felis rufa
Lynx
 Felis lynx

Coyote
 Canis latrans
Marmot
 Marmota spp.
Pika
 Ochotoma collaris
Mink
 Mustela vison
Beaver
 Castor fiber
Great Blue Heron
 Ardea herodias
Trumpeter Swan
 Cygnus buccinator
Red-tailed Hawk
 Buteo jamaicensis
Golden Eagle
 Aquila chrysaetos
Bald Eagle
 Haliaetus leucocephalus
Blue Grouse
 Dendragapus obscurus
Sandhill Crane
 Grus canadensis

Visiting: There are organized tours of the area and visitors may motor the Park roads. Horse and boat tours are limited and foot packing is allowed under controlled conditions.

189

Index

Picture Credits

Jacket: M. D. England (*Ardea*)

Ardea: 46, 48-49 (top), 54, 63, 107,
139, 179; Peter Alden—43 (top),
101; Gert Behrens—67; Hans &
Judy Beste—18 (top), 19, 22-23,
30-33, 58-59, 111, 119 (top and
bottom); P. Blasdale—65, 184;
R. M. Bloomfield—66, 135, 146,
149, 183; J. B. & S. Bottomley—17
(top), 116 (bottom), 124, 153
(bottom); Elizabeth S. Burgess—
41; J. J. Buxton—172; Kevin
Carlson—57, 62; Graeme
Chapman—89, 112; F. Collet—
60-61 (top); Werner Curth—55;
M. D. England—56, 91, 100, 115,
130-132, 133 (bottom), 137 (top
and bottom), 141 (bottom), 154,
157, 158-160, 175, 178; Kenneth
W. Fink—16, 17 (bottom), 25 (top),
42, 70, 72, 80-81, 102, 105, 122-123,
153 (top), 174, 186-188; Clem
Hanger—147; Peter Johnson—
168-169; Chris Knights—37
(bottom); M. Kirshnan—48
(bottom); Ake Lindau—37 (top),
116 (top), 117, 177; Eric Lindgren–
110; Edwin Mickleburgh—10-11,
13; P. Morris—2-3, 40, 73, 103;
S. Roberts—129 (top); Bryan L.
Sage—29 (top and bottom), 125,
151; Robert T. Smith—171; Peter
Steyn—126; Bernard Stonehouse—
12; W. Stribling—51 (top and
bottom); Valerie Taylor—86-87,
167; W. R. Taylor—18 (bottom);
Richard Vaughan—34-35, 127, 141
(top); Richard Waller—47, 53,
93-99, 155 (bottom), 161; Adrian
Warren—45, 71-75, 85;
C. Weaver—43 (bottom), 90, 142,
165; Alan Weaving—66-67, 106,
109, 121 (top), 145, 155 (top), 164,
182, 185; Tom Willock—78, 81
(right), 189; J. S. Wightman—26-
27, 83, 121 (bottom), 129
(bottom), 133 (top), 134, 143, 173,
181.
*Australian News and Information
Service*: Don Edwards—20-21;
H. Frauca—61 (bottom).
National Audubon Society; 25
(bottom).
Survival Anglia—162-163.
*United States National Park
Service*—39.
*United States Fish and Wildlife
Service*—14-15.